COMBINING GREEN-BLUE-GREY INFRASTRUCTURE FOR FLOOD MITIGATION AND ENHANCEMENT OF CO-BENEFITS

Alida I. Alves Beloqui

COMBINING GREEN-BLUE-GREY INFRASTRUCTURE FOR FLOOD MITIGATION AND ENHANCEMENT OF CO-BENEFITS

DISSERTATION

Submitted in fulfillment of the requirements of

the Board for Doctorates of Delft University of Technology

and

of the Academic Board of the IHE Delft

Institute for Water Education

for

the Degree of DOCTOR

to be defended in public on

Thursday, 30 January 2020, at 15.00 hours

in Delft, the Netherlands

by

Alida Ivana ALVES BELOQUI

Master of Science in Municipal Water and Infrastructure, IHE Delft, the Netherlands
and Asian Institute of Technology, Thailand

born in Montevideo, Uruguay

This dissertation has been approved by the
promotor: Prof.dr. D. Brdjanovic and
copromotor: Dr. Z. Vojinovic

Composition of the doctoral committee:

Rector Magnificus TU Delft	Chairman
Rector IHE Delft	Vice-Chairman
Prof.dr. D. Brdjanovic	IHE Delft / TU Delft, promotor
Dr. Z. Vojinovic	IHE Delft, copromotor

Independent members:

Prof.dr.ir. A.E. Mynett	TU Delft / IHE Delft
Prof.dr. D.A. Savic	University of Exeter, United Kingdom
Prof.Dr.-Ing. P. Fröhle	Hamburg University of Technology, Germany
Dr.ir. F.H.M. van de Ven	TU Delft
Prof.dr. M.J. Franca	TU Delft / IHE Delft, reserve member

This research was conducted under the auspices of the SENSE Research School for Socio-Economic and Natural Sciences of the Environment

CRC Press/Balkema is an imprint of the Taylor & Francis Group, an informa business

© 2020, Alida Ivana Alves Beloqui

Published by:
CRC Press/Balkema
Schipholweg 107C, 2316 XC, Leiden, the Netherlands
Pub.NL@taylorandfrancis.com
www.crcpress.com – www.taylorandfrancis.com
ISBN: 978-0-367-48597-9

SUMMARY

Climate change is presenting one of the main challenges to our planet. In parallel, all regions of the world are projected to urbanise further. Consequently, sustainable development challenges will be increasingly concentrated in cities. A resulting impact is the increment of urban flood risk in many areas around the globe, where it is expected to have higher flooding frequency in the future. In these cases, appropriate flood risk management is crucial, but decision-makers face a big challenge due to the complexity of urban drainage systems. These systems are complex due to the number of possible measures, the need for high investments and the uncertainty about future conditions. Conventional approaches to flood risk management, which frequently do not address the root causes of risk, are based on grey infrastructure. A change of paradigm is needed to develop effective adaptation strategies for current and future scenarios.

Green-blue infrastructure (GBI) is a central concept to achieve adaptation to the effects of climate change. Its main strength is the ability to deliver environmental, social and economic benefits simultaneously. Meanwhile, grey approaches are mostly single-objective oriented designs and frequently present conflicting interests in urban spaces. Currently, the combination of green and grey measures, also called 'hybrid' approaches, is being suggested. As a result of this combination, grey measures can be complemented by GBI since each approach has its own advantages and meets different goals.

Even though strong evidence exists demonstrating the effectiveness of GBI as a sustainable solution to reduce urban flooding, its adoption by cities is still slow. The implementation of GBI has important obstacles, such as lack of technical references, low public acceptability, lack of stakeholder collaboration and uncertainty about performance and costs. A crucial element to increase application of GBI is the emphasis on the provision of multiple benefits in addition to flood protection. Moreover, the economic analysis of these co-benefits can have a significant impact on decision-making proving that the investment in climate change adaptation using GBI is economically efficient.

The general objective of this research is to contribute to decision-making processes for selection and assessment of adaptation strategies to cope with urban flood risk while achieving other benefits. With this aim, several specific objectives were defined: to develop a framework to introduce co-benefits into decision making processes for stormwater infrastructure planning; to examine how preferences regarding key benefits differ among different stakeholders; to assess how the quantification of the multiple benefits of GBI can help to justify its implementation; to evaluate the effects of including co-benefits on the selection of flood risk reduction strategies and assess the trade-offs among cost and benefits.

In this work a multi-criteria method for measures screening has been developed and tested. It allows selecting among different types of measures to reduce different types of flood risk. Three goals were considered to assess measures performance from a holistic point of view: flood risk reduction, cost minimization and co-benefits enhancement. This method offers a useful preliminary analysis to facilitate the application of more complex and time-consuming evaluation frameworks, such as hydrodynamic modelling and optimisation techniques.

This research also analyses preferred co-benefits among different types of stakeholders, allowing to visualise the importance of including all relevant actors when making decisions to plan flood risk infrastructures. In particular, contributions from local stakeholders help to better understand local conditions and needs. The results obtained show the importance of participatory planning processes, confirming that unilateral planning processes led by policy makers or scientific community, could focus on benefits that are not relevant for local residents and reduce measures' acceptability.

This work also focuses on including co-benefits quantification into the cost-efficiency assessment of measures. To achieve this, a monetary analysis of co-benefits is introduced into a cost-benefit analysis of flood risk mitigation measures. The results obtained illustrate in quantitative terms how the viability of GBI for flood mitigation is considerably improved when co-benefits are considered. Thus, it is important to consider co-benefits when planning adaptation strategies to improve urban flood risk management, otherwise GBI is likely to appear less efficient than grey infrastructure. The results also showed that a mix of green-blue and grey infrastructures is likely to result in the best adaptation strategy as they tend to complement each other. While grey infrastructure is good at reducing the risk of flooding, green-blue infrastructure brings in multiple additional benefits that grey infrastructure cannot deliver.

Lastly, the valuation of co-benefits is integrated into an optimisation framework. The results obtained confirm optimisation as a helpful decision-making tool for flood risk management. Especially, it allows comparing among optimal combinations of green-blue and grey measures for a wide range of costs. An analysis of how the effectiveness of optimal solutions regarding the primary function of flood risk reduction varies when the objectives are altered is also provided. This allows visualizing in a quantitative way the trade-offs when multiple benefits are pursued in flood risk infrastructure planning. The approach shows that there are inevitable trade-offs among the benefits obtained from green-blue and grey measures.

This research contributes to improving planning processes for flood risk management in urban spaces. Current perspectives call for infrastructures which can integrate multiple challenges at the same time, in order to create safe and liveable urban spaces for current and future conditions. Thus, this research provides tools and knowledge to facilitate holistic decision making.

SAMENVATTING

Klimaatverandering is een van de grootste uitdagingen voor onze planeet. Tegelijkertijd wordt verwacht dat alle regio's van de wereld verder zullen verstedelijken. Daarom zullen de uitdagingen op het gebied van duurzame ontwikkeling vooral in steden geconcentreerd zijn. Een gevolg hiervan is een toegenomen overstromingsrisico in stedelijke gebieden over de hele wereld, waar verwacht wordt dat er vaker overstromingen zullen zijn in de toekomst. Daarom is goed overstromingsrisico management cruciaal in stedelijk gebied, maar beslissingsmakers hebben te maken met grote uitdagen vanwege de complexiteit van stedelijke afwateringssystemen. Deze complexiteit stamt uit de hoeveelheid mogelijke maatregelen, de hoge investeringen en de onzekerheid over toekomstige situaties. Conventionele aanpakken van overstromingsrisico management zijn gebaseerd op grijze infrastructuur, die vaak niet de oorzaak van de risico's adresseert. Een paradigmaverandering is nodig om effectieve adaptatie strategieën te ontwikkelen voor huidige en toekomstige scenario's.

Groen-blauwe infrastructuur (GBI) is een centraal concept om adaptatie aan de effecten van klimaatverandering te bereiken. Haar belangrijkste sterkte punt is de mogelijkheid om simultaan milieu -, sociale - en economische voordelen te bieden. Terwijl grijze aanpakken meestal ontworpen zijn om één doel te bereiken, en vaak tegengestelde belangen in de stedelijke omgeving creëren. Momenteel wordt de combinatie van groene en grijze maatregelen, ook wel "hybride" aanpakken genoemd, voorgesteld. In deze combinatie worden grijze maatregelen aangevuld met GBI omdat elke aanpak zijn eigen voordelen heeft en verschillende doelen bereikt.

Alhoewel er sterk bewijs is voor de effectiviteit van GBI als een duurzame oplossing om overstroming van stedelijke gebieden te voorkomen, wordt het nog weinig gebruikt door steden. Er zijn belangrijke obstakels voor de implementatie van GBI, zoals het gebrek aan technische referenties, lage publieke acceptatie, gebrek aan samenwerking tussen de verschillende stakeholders en onzekerheid over de prestaties en kosten. Een cruciaal element om het gebruik van GBI te doen toenemen is de nadruk op het feit dat GBI meer voordelen biedt naast bescherming tegen overstroming. De economische analyse van deze bij-voordelen kan een significante impact hebben op de besluitvorming, door te bewijzen dat de investering in klimaatadaptatie doormiddel van GBI economisch efficiënt is.

Het doel van dit onderzoek in algemene zin is om bij te dragen aan het besluitvormingsproces voor de selectie en assessment van adaptatie strategieën om om te gaan met stedelijk overstromingsrisico, terwijl er ook andere voordelen worden bereikt. Naast dit doel zijn verschillende specifieke doelstellingen gedefinieerd: het ontwikkelen van een framework dat de bij-voordelen meeweegt in het besluitvormingsproces rondom stormwater infrastructuur planning; onderzoeken hoe de voorkeuren voor belangrijkste voordelen verschillen tussen de verschillende stakeholders; onderzoeken hoe de kwantificering van de verschillende voordelen van GBI kunnen helpen om de implementatie te rechtvaardigen; evalueren hoe de effecten van het meewegen van bij-voordelen op de selectie van overstromingsrisico reductie strategieën en de afweging tussen kosten en baten.

In dit werk is een multi-criteria methode voor het screenen van maatregelen ontwikkeld en getest. Daarin kunnen verschillende typen maatregelen om verschillende typen overstromingsrisico te verminderen worden geselecteerd. Drie doelen zijn opgenomen om de uitkomsten van de maatregelen vanuit een holistisch perspectief te bekijken: overstromingsrisico reductie, kostenbeheersing en het vergroten van de bij-voordelen. Deze methode biedt een bruikbare preliminaire analyse om de toepassing van meer complexe en tijdrovende evaluatie frameworks te faciliteren, zoals hydrodynamisch modelleren en optimalisatie technieken.

Dit onderzoek analyseert ook de bij-voordelen die de voorkeur hebben bij verschillende typen stakeholders, waardoor het mogelijk wordt om te visualiseren wat het belang is van het meenemen van alle relevante actoren wanneer er beslissingen worden genomen om overstromingsrisico infrastructuur te plannen. De bijdrage van lokale stakeholders helpen in het bijzonder om lokale condities en behoeften te begrijpen. De resultaten die zijn verkregen laten het belang van een participatief planningsproces zien, en bevestigen dat unilaterale planningsprocessen, geleid door beleidsmakers of wetenschappers kunnen focussen op voordelen die niet relevant zijn voor lokale bewoners, en de acceptatie van de maatregelen verkleint.

Dit onderzoek focust ook op het meewegen van de kwantificering van de bij- voordelen in het onderzoek naar de kosteneffectiviteit van de maatregelen. Om dit te bereiken, is er een monetaire analyse van de bij-voordelen geïntroduceerd in de kosten-baten analyse van overstromingsrisico mitigatie maatregelen. De verkregen resultaten illustreren in kwantitatieve termen hoe de levensvatbaarheid van GBI voor overstromingsrisico mitigatie verbetert wanneer bij-voordelen worden meegewogen. Daarom is het belangrijk om bij-voordelen mee te wegen wanneer adaptatie strategieën om stedelijk overstromingsrisicomanagement te verbeteren, anders is het waarschijnlijk dat GBI minder efficiënt lijkt dan grijze infrastructuur. De resultaten lieten ook zien dat een mix van groen-blauwe en grijze infrastructuur waarschijnlijk zal resulteren in de beste adaptatie strategie, omdat ze elkaar over het algemeen aanvullen. Waar grijze infrastructuur goed is in het verminderen van overstromingsrisico, brengt grijs-blauwe infrastructuur meerdere toegevoegde voordelen die grijze infrastructuur niet kan bieden.

Ten laatste is de waardering van bij-voordelen geïntegreerd in een optimalisatie framework. De verkregen resultaten bevestigen dat optimalisatie een waardevolle besluitvormingstool is voor overstromingsrisicomanagement. Het zorgt ervoor dat de optimale combinaties van groen-blauwe en grijze maatregelen voor een groot bereik aan kosten kunnen worden vergeleken. Een analyse van hoe de effectiviteit van optimale oplossingen betreffende de primaire functie van overstromingsrisico reductie varieert wanneer de doelen worden verandert wordt ook geleverd. Dit zorgt ervoor dat op een kwantitatieve manier de afwegingen kunnen worden gevisualiseerd wanneer verschillende voordelen worden nagestreefd bij overstromingsrisico infrastructuur planning, en het laat zien dat er onafwendbare trade-offs zijn tussen de voordelen van groen-blauwe en grijze maatregelen.

Dit onderzoek draagt bij aan de verbetering van het planningsproces van overstromingsrisico management in stedelijke gebieden. De huidige perspectieven vragen om infrastructuur die verschillende uitdagingen kan integreren, om veilige en leefbare stedelijke gebieden voor huidige en toekomstige condities te creëren. Daarom is holistische besluitvorming nodig en dit onderzoek draagt tools en kennis bij om dat te faciliteren.

CONTENTS

1

INTRODUCTION

1.1 PROBLEM STATEMENT

1.1.1 Main drivers for urban adaptation

Climate change is presenting one of the main challenges to our planet. Urban areas are expected to suffer more from the impacts of climate change compared to other landscapes, with more frequent extreme weather events (EEA, 2016; Kabisch *et al.*, 2017a). In parallel to climate change, there is an ongoing global urbanisation process. Already half of the world population lives in urban areas and over the coming decades all regions of the world are projected to urbanise further. By 2050 around 66% of the global population is expected to live in cities. Consequently, sustainable development challenges will be increasingly concentrated in cities (United Nations, 2014).

The impacts of climate change on society comprise health-related and socio-economic problems originated by increasingly frequent heatwaves, droughts and flooding events (EEA, 2016). A main impact in urban spaces is the increase of flood risk. Floods are one of the most frequent and damaging natural disasters; more than 90% of floods have meteorological causes and most of them have their origin in heavy rainfall (Simonovic, 2012; World Bank, 2017). Flooding is a global phenomenon which is causing enormous economic damages and loss of human lives. Economic losses from weather and climate-related disasters have increased in the last 30 years and urban areas have been particularly impacted recently (IPCC, 2012; Jha *et al.*, 2012).

Due to climatic changes, the frequency of heavy rainfall is expected to increase in many regions, increasing the frequency of floods (IPCC, 2012). In addition, the replacement of vegetation with artificial surfaces due to urbanisation is decreasing surface permeability, generating more runoff which will impact negatively on the performance of drainage systems (EEA, 2012). The combined effect of these drivers and the trend from them suggests an important increase of future flood risk (EEA, 2012; IPCC, 2012). This indicates the need for rapid action and changes in how urban drainage systems are planned (Zhou *et al.*, 2019). The development of sustainable management approaches which integrate mitigation and adaptation strategies is becoming increasingly important (Vojinovic, 2015). These drainage strategies should co-optimize flood risk reduction with other objectives to ensure habitable cities (Arnbjerg-Nielsen *et al.*, 2013; Yazdanfar and Sharma, 2015).

Effective flood risk management is crucial to protect people and mitigate future damage. However, cities are complex systems which integrate social, ecological, and technical aspects. The combination of these characteristics with the impacts of climate change and population growth is generating big challenges for decision-makers (McPhearson *et al.*, 2016). Besides, urban drainage systems have their own complexity: costly rehabilitation of existent systems, the quantity of available measures, the significant investments needed to implement these strategies, and the uncertainty about future conditions, are increasing decision-making complexity even more (Chocat *et al.*, 2007; Jha *et al.*, 2012; Simonovic, 2012).

A change of paradigm is needed, enabling adaptation to climate extremes in the long term through the enhancement of social welfare, quality of life, infrastructure, and the incorporation of a multi-hazard approach in planning for disasters in the short term (IPCC, 2012). To develop effective adaptation strategies for current and future scenarios it is important to incorporate the use of green-blue measures, which should be evaluated analysing their applicability in different contexts and their potential to achieve multiple benefits related with urban well-being, rather than just the management of stormwater (Vojinovic, 2015).

1.1.2 Green-blue infrastructure for adaptation

Decision-makers carrying out urban planning and infrastructure development processes have to take into account how to reduce cities' contribution to climate change (mitigation) as well as the vulnerability to the impacts of climate change (adaptation). Even though climate change adaptation and mitigation is crucial, local adaptation measures are more effective for flood risk management than global mitigation programs (Zhou *et al.*, 2018). Moreover, urban sustainability needs to be considered during decision-making processes. This means to achieve development that improves the quality of life for citizens in the present while safeguarding the wellbeing of future generations (Elmqvist *et al.*, 2019). This can be accomplished by seeing climate-change adaptation as an opportunity to create innovative solutions alongside traditional measures, ensuring more attractive and safe cities currently and in the future, through the provision of multiple benefits (EEA, 2012; 2016).

In the area of urban drainage management, similar concepts are named with different terms in different parts of the world. This growth of urban drainage terminology is the consequence of the growing interest in urban stormwater management in the last few decades. Terms such as BMPs (best management practices), LIDs (low impact development), WSUD (water sensitive urban design), SuDS (sustainable drainage systems), GBI (green-blue infrastructure), EbA (ecosystem-based adaptation) and NBS (nature-based solutions) are broadly used. This variety of terms and concepts brings about the possibility of miscommunication, which should be minimised through the careful use of terminology (Fletcher *et al.*, 2014).

Green infrastructure is defined as an interconnected network of multifunctional green spaces which together maintain and enhance ecosystem services and resilience, providing multiple functions and services to people, the economy and the environment (Tzoulas *et al.*, 2007; Naumann *et al.*, 2011; European Commission, 2012a). Nature-based solutions (NBS) is a relatively new concept defined as solutions inspired and supported by nature which offer multiple benefits helping society to mitigate and adapt to climate change (Nesshöver *et al.*, 2017; Raymond *et al.*, 2017). Dorst *et al.* (2019) compare NBS principles with those of GI, finding several similarities: both concepts refer to interventions based on nature which provide multi-functionality, addressing social, economic and environmental challenges simultaneously, and involving holistic and participatory planning. Even though the authors stress that the concept of NBS includes a wider range of interventions and non-natural technology, GBI is seen as a subset of NBS and they can sometimes be synonymous (Nesshöver *et al.*, 2017; Lafortezza *et al.*, 2018). Through this work the terms GBI, NBS, non-traditional measures and sustainable solutions are used interchangeably, referring to the concept of measures or solutions based in nature or natural processes.

GBI is a central concept to achieve adaptation to the effects of climate change since it includes measures or actions to reduce society's vulnerability and to expand resilience capacity (IPPC, 2007). This is based on the ability of GBI to deliver multiple services, providing environmental and cultural benefits and at the same time contributing to climate change adaptation and mitigation (Kabisch *et al.*, 2016). Therefore, GBI can create win-win situations, and for this reason several cities are using these measures as cost-effective and integrated climate adaptation solutions (Liu and Jensen, 2018; Miller and Montalto, 2019).

Different works have shown the effectiveness of these infrastructures in mitigating flood risk (Kong *et al.*, 2017; Zölch *et al.*, 2017; Versini *et al.*, 2018), performing in a similar way to grey infrastructure, with comparable cost and additional benefits (Liquete *et al.*, 2016). These measures reduce flood risk by means of mitigating physical vulnerability to the hazard of flooding (Cardona *et al.*, 2012; Vojinovic *et al.*, 2016a). Moreover, since GBI can provide several related co-benefits, these are options that provide benefits even in the absence of climate change, consequently seen as 'low-regret' solutions (EEA, 2012; Casal-Campos *et al.*, 2015).

Meanwhile, grey or conventional approaches in the case of flood risk management frequently do not address the root causes of risk and can even raise the vulnerability of populations in the long term. Conventional drainage solutions are focused on efficient collection and fast conveyance of water through piped systems or underground storage (USEPA, 2000). These are mostly single-objective oriented designs, with high cost, low flexibility and frequent conflicting interests in the urban space (Brink *et al.*, 2016). Although such approaches have diminished flooding damage during the previous two centuries and are still necessary to cope with extreme flood events, sustainable alternatives which offer additional benefits are increasingly being favoured (Kabisch *et al.*, 2017a).

'Hybrid' approaches, combining GBI and grey infrastructure, seems to be the most effective strategy in an urban context to mitigate flooding hazards and enhance system resilience (Kabisch *et al.*, 2017a; Xie *et al.*, 2017; Haghighatafshar *et al.*, 2018; Browder *et al.*, 2019). Conventional grey approaches can be complemented by GBI since each approach has its own advantages and meets different goals (Dong *et al.*, 2017). This can lead to a new generation of solutions to enhance the performance of drainage systems and protect communities (Browder *et al.*, 2019). Moreover, solutions combining grey and green infrastructures are likely more robust and flexible over a long period of time (EEA, 2012).

Even though strong evidence exists demonstrating the effectiveness of GBI as a sustainable solution to reduce urban flooding and complement existing drainage systems, these measures are still being applied at a slow pace in cities (Dhakal and Chevalier, 2017; Qiao *et al.*, 2018). Further actions are needed to demonstrate and communicate the full potential of GBI, as well as to increase its acceptance and application (Kabisch *et al.*, 2017a).

1.1.3 Barriers for GBI implementation

Traditional grey infrastructure continues to be widely preferred in urban areas throughout the world (Dhakal and Chevalier, 2017). The implementation of GBI has important obstacles due to knowledge gaps and the need to involve different stakeholders and disciplines. Several barriers for GBI acceptance are identified. From a technological point of view, while traditional approaches count on enough technical support and tools for decision making, GBI for stormwater management lacks sufficient technical references, standards and guidelines (Qiao *et al.*, 2018). In particular, this support is lacking regarding the evaluation and quantification of additional benefits (IPCC, 2012). Another commonly identified barrier is uncertainty about long-term performance and cost-effectiveness compared to conventional solutions; this has a major effect on GBI acceptance and implementation (Davis *et al.*, 2015). Furthermore, institutional barriers, public acceptability and lack of stakeholder collaboration are also identified as barriers for GBI application (Haghighatafshar *et al.*, 2018; Liu and Jensen, 2018; Qiao *et al.*, 2018; Thorne *et al.*, 2018; Wihlborg *et al.*, 2019).

The complexity of problems and the variety of alternatives requires careful selection of adaptation measures identifying impacting factors for every local area to establish location-based solutions (Yazdanfar and Sharma, 2015). Besides, the site-specific nature of GBI requires these measures to be designed for each case individually, restricting the development of standard solutions, which implies one extra step for the difficulty of measure selection and design. Moreover, because of this site-specific nature, the levels of effectiveness and costs vary significantly from case to case increasing uncertainties for decision makers (Davis *et al.*, 2015).

The integration of goals from different policy sectors and the involvement of a wide range of stakeholders, combining multiple and frequently conflicting interests, are crucial to raising GBI application and acceptance (Davis *et al.*, 2015; Hoang and Fenner, 2016; Raymond *et al.*, 2017). However, this often implies that different disciplines or departments work together which can be an institutional issue (Matthews *et al.*, 2015). Problems related to fragmentation and low stakeholder participation are described by Brink *et al.* (2016), who found out that most research about GBI focuses on solving heat or flooding issues, while economic and social evaluations are scarce. Besides, they found that most works about adaptation at the municipal level do not include municipal participation, suggesting a gap between research and practice. Aligned with this, Bissonnette *et al.* (2018) argue that the slow application of GBI is not based on technical issues, rather it originates in the lack of collaboration during the design and implementation of GBI. By exploring the interests of different stakeholders, we can better understand drivers and barriers for water system transformation, in particular regarding the implementation of GBI for water management (Albert *et al.*, 2019).

An example about this is given by Miller and Montalto (2019), who performed stakeholder surveys in NYC concluding that even if the primary driver of GBI application is stormwater management, local residents perceive other ecosystem services more positively. As a result, multifunctional GBI programs may have better public support than the ones concentrated solely on stormwater management. Also, the work done by Derkzen *et al.* (2017) concludes that citizens are willing to support multifunctional GBI, for example providing recreational and aesthetic benefits. The authors suggest the delivery of information about the multiple benefits of GBI to achieve public support, and in addition they advise that the GBI choice is adapted to local preferences.

Therefore, further actions are needed to increase the acceptance of GBI over grey infrastructure for water management. To achieve this, the emphasis on the provision of multiple benefits in addition to flood protection is a crucial element (Kabisch et al., 2017a).

1.1.4 The multiple benefits of GBI

GBI offers a multifunctional, solution-oriented approach to enhance urban sustainability (Dorst *et al.*, 2019). The simultaneous delivery of social, economic and environmental benefits by GBI increases the willingness to accept and implement these solutions. Awareness about the co-benefits provided by GBI and its economic assessment can be crucial to convince decision-makers about the feasibility and necessity of GBI implementation (EEA, 2012; Liu and Jensen, 2018; Qiao *et al.*, 2018). Besides, when designing GBI for stormwater management, these co-benefits should be identified from the beginning of the planning process and according to the specific location needs. This approach will result in better co-design in which flood management functions as well as co-benefits are equally ensured among the goals pursued and not left to occur accidentally (Kremer *et al.*, 2016; Fenner, 2017).

In addition to flood protection, GBI provides multiple secondary benefits. When working on flood risk management, the reduction of flood damage is the primary benefit, while the secondary benefits are also called *co-benefits* (i.e. Dagenais *et al.*, 2017). Examples of these co-benefits are: CO_2 storage capacity, increasing biodiversity, improvements in public health, and recreation opportunities. The net benefits obtained from the application of GBI can significantly exceed those of conventional solutions if co-benefits and a long-term perspective are considered. An example of this is given by a study from the stormwater management programme in the city of Philadelphia, USA. This study established that the net benefits of using surface measures were almost 30 times higher than the benefits obtained from the piped alternative (Davis *et al.*, 2015).

The economic analysis of these co-benefits can have a significant impact on decision-making by proving that an investment in climate change adaptation using GBI is economically efficient. Moreover, economics of adaptation strategies are an important component of decision-making because it establishes evidence-based decisions and allows its financial consequences to be visualized (EEA, 2016). However, there are also co-benefits of GBI which are not easily represented in monetary terms. Since it is important to consider the whole range of benefits provided, efforts should be made to take into consideration non-monetary criteria to include benefits such as health enhancement, human well-being, liveability improvements, and conservation of natural resources (EEA, 2016; Kabisch *et al.*, 2017b).

Even though the estimation of complete benefits of applying GBI is challenging, decision-makers prefer quantitative data for their decisions, thus the acceptance of these measures can be reinforced by making these solutions financially attractive (Stratus Consulting, 2009; Machac *et al.*, 2018). Cost–benefit analysis can help to predict if the benefits from an adaptation measure outweigh the costs, but it is an economic evaluation method in which all costs and benefits need to be expressed in monetary terms (Saarikoski *et al.*, 2016).

Zölch *et al.* (2018) examined municipal climate adaptation strategies in Germany. They observed a growing acknowledgement of the improvement of ecosystem services when implementing urban adaptation plans, since three quarters of the analysed strategies mentioned at least some type of GBI. However, the ratio of GBI to conventional adaptation alternatives was often low and only 25% of cases emphasized the multiple benefits of GBI, suggesting that this aspect is not among the implementation criteria. Hence, the acceptance of GBI as an adaptation option needs more research to evaluate both the benefits and the cost-effectiveness of GBI compared to conventional options.

To summarize, in order to develop effective long-term adaptation strategies, it is important to analyse the applicability of measures in each context. Moreover, the measures have to be evaluated taking into account multiple challenges, local preferences involving different stakeholders, as well as considering GBI potential to achieve multiple benefits (Vojinovic, 2015; Raymond et al., 2017). There is a need for research to establish methods to determine the most efficient infrastructure combination, while optimizing the multiple benefits and discovering the trade-offs that may be generated by GBI involving economic, ecological and social benefits (Kabisch *et al.*, 2017a).

1.2 RESEARCH THEME AND OBJECTIVES

1.2.1 Thesis focus

So far several arguments describing the motivation of this work have been presented. Summarising:

- A change of paradigm is needed to cope with current and future urban challenges, in particular to mitigate flood risk in a sustainable way.

- Although GBI has been recognised as a promising approach to achieve this, several barriers prevent a broad application of these measures in urban spaces.

- The analysis of the multiple benefits provided by GBI and in particular their monetary valuation, has been suggested as an effective approach to overcome this problem, encouraging decision makers to choose GBI.

The aim of this work is to reduce the barriers of GBI implementation for flood mitigation in urban spaces. To achieve this goal, this thesis focuses on improving our knowledge about how to include the multiple benefits of GBI into the selection and planning of urban drainage infrastructure. A better understanding of these elements will help decision makers when planning integrated and sustainable urban drainage infrastructures which help to efficiently cope with multiple challenges at the same time. Next, the decision making steps on which this research has impact are identified.

Decision making regarding water resources management consists of several steps (Figure 1.1), all of which are crucial to reach effective solutions (Department for Communities and Local Government, 2009; Simonovic, 2009). The first step is to define the problem, which is based on a risk assessment to identify priority intervention areas and elements of the urban system. The second step is to indentify applicable intervention options or strategies that contribute to the achievement of the objectives defined in the previous step. These options attempt to improve general adaptive capacity, for instance reducing climate risk to an acceptable level. These strategies comprise grey or structural, green and soft options, such as early warning systems (EEA, 2012). Third, it is necessary to establish criteria for evaluating the alternatives. These criteria will allow the comparison among options.

In a fourth step, after possible adaptation strategies have been identified, an assessment is needed to determine which of them are better for the specific case. The assessment should be based on their feasibility and their capacity to reach the adaptation goals. Next, the best strategy is chosen; this strategy has to be effective and efficient. Effective strategies reduce vulnerabilities to an acceptable level. Efficient strategies have benefits (economic, social and environmental) higher than their costs and are more cost-effective than other options (EEA, 2012). Afterwards, the chosen strategy is designed and then implementation and operation plans are defined (Simonovic, 2009). Finally, good decision making implies a continuous re-examination of past choices, this allows to learn from previous mistakes in order to inform future decisions (Department for Communities and Local Government, 2009).

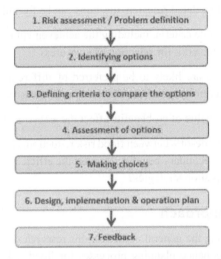

Figure 1.1. Steps for decision making in flood risk management; steps 2 to 5 (in orange) are the scope of this thesis.

This research contributes to four of these steps: the identification of applicable options, the definition of assessment criteria, the evaluation of alternatives and the selection of the most adequate strategy. Specifically, this research concentrates on adaptation strategies oriented to solve multiple challenges simultaneously, maximising multiple benefits, considering local needs and integrating the point of view of diverse stakeholders.

1.2.2 Objectives

The general objective of this research is to contribute to the decision-making processes for selection of adaptation strategies to cope with urban flood risk while achieving other benefits. Specific objectives are:

- to develop a framework for flood mitigation infrastructure selection taking into account multiple benefits;

9

- to better understand how preferences and perceptions of key benefits may differ among different stakeholders and how can they influence planning decisions;

- to assess how the quantification of the co-benefits can influence decisions about flood mitigation measures implementation;

- to evaluate how the incorporation of co-benefits within economic evaluation of flood mitigation measures can affect the trade-offs among cost and benefits.

1.2.3 Research questions

Given the objectives, the related research questions can be formulated as:

RQ1: How can multiple criteria, including the achievement of secondary benefits, be integrated into a framework for selection of measures?

RQ2: Which benefits are likely to be preferred by different groups of stakeholders and how can these preferences affect the selection of measures?

RQ3: How does the value of co-benefits affect the assessment of adaptation options?

RQ4: What are the trade-offs between flood risk reduction, cost and co-benefits? Does the enhancement of secondary benefits decrease the efficiency of flood risk reduction when selecting adaptation strategies?

1.2.4 Research approach

With the aim of fulfilling the general objective, this research concentrates on providing several elements that enhance planning processes for flood risk management in urban environments. Current perspectives call for infrastructures which can integrate multiple challenges at the same time, in order to create safe and liveable urban spaces for current and future conditions. Thus, holistic decision making is needed and this research provides tools and knowledge to facilitate its achievement. This work is divided into four parts, oriented to address each research question.

An initial screening of measures before going into complex assessment of strategies is needed, and this screening should be done integrating several elements, such as combination of green-blue and grey options, multiple benefits and local characteristics. In this work multi-criteria analysis is applied to achieve this.

The co-benefits provided by GBI are numerous and different measures perform differently in the achievement of them. With the objective of improving the selection of measures according to site site-specific needs, it is important to identify the preferred co-benefits of different stakeholders. In this work a survey is applied to improve this understanding.

The advantages of considering multiple benefits have already been discussed. The inclusion of these benefits in life cycle cost-benefit analysis can have a significant impact on decision making regarding GBI. This research compares the results of economic analysis of flood risk strategies with and without co-benefits.

Lastly, after verifying that the inclusion of co-benefits in decision-making processes stimulates the selection of GBI above grey measures, it seems important to evaluate the trade-offs among cost and benefits when different measures are applied. In this study hydrodynamic models and optimisation techniques are combined for this assessment.

1.2.5 Thesis outline

This thesis is structured in six chapters (Figure 1.2). Chapter 1 provides the theoretical background and describes the objectives and structure of the thesis. The next four chapters are focused on the four specific objectives previously presented. Each one of these four chapters is based on a peer-reviewed publication, published or in the process of being published. Figure 1.2 shows the different chapters, interlinks among them embedded in the general planning process, and how they are related with the research questions.

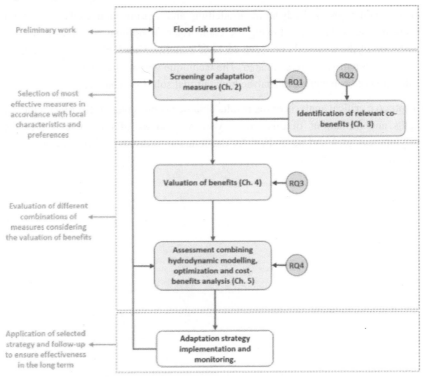

Figure 1.2. Overview of methodological approach, chapter contributions and their link with each research question (RQ)

Chapter 2 presents a multi-criteria framework for the selection of measures. This framework integrates the co-benefits besides flood characteristics and local physical features. The framework was applied in three study areas; in each case decision makers followed simple multiple-choice steps to develop a ranking of green-blue and grey solutions.

Chapter 3 highlights two main steps in stormwater strategy selection considering co-benefits. The first part involves a description of co-benefits that can be obtained from GBI, providing a link between urban ecosystem services and GBI. Secondly, the method focuses on the analysis of stakeholders' preferences regarding co-benefits as a main input for the selection of measures. Three stakeholders' groups are considered: the general public, policy makers and the scientific community.

Chapter 4 presents a method for the valuation of flood damage reduction (primary benefit) and co-benefits. These values are included in a lifespan cost-benefit analysis to compare different combinations of green-blue-grey measures. Rankings of measures are built considering different objectives. Results vary when the focus is shifted from improving only the primary benefit to achieving total benefits (primary benefit plus co-benefits).

Chapter 5 combines hydrodynamic modelling and optimisation tools as a helpful decision-making technique, allowing the comparison of optimal strategies and a clear visualization of the trade-offs between cost and benefits. Green-blue and grey measures and their combinations were evaluated. The technique was applied with and without the consideration of co-benefits, to analyse how optimal solutions change.

In Chapter 6 a critical analysis of the results obtained is presented, reflecting on the main strengths and limitations of this work. Also, an outlook of this topic is presented, identifying new gaps and necessities for the future in order to further contribute to the development of this subject.

2

MULTI-CRITERIA APPROACH FOR SELECTION OF GREEN-BLUE AND GREY INFRASTRUCTURE

Continuous changes in climate conditions combined with urban population growth pose cities as one of the most vulnerable areas to increasing flood risk. In such an atmosphere of growing uncertainty, a more effective flood risk management is becoming crucial. Nevertheless, decision-making and selection of adequate systems is a difficult task due to complex interactions between natural, social and built environments. The combination of green-bue (or sustainable) and grey (or traditional) options has been proposed as a way forward to ensure resilience in advance of extreme events, and at the same time to obtain co-benefits for society and the environment. This chapter describes a method for selection of urban flood measures, based on a multi-criteria analysis that includes flood risk reduction, cost minimization and enhancement of co-benefits. The aim of this method is to assist decision makers in selecting and planning measures, which afterwards can be part of either high level scoping analysis or more complex studies, such as model based assessment. The proposed method is implemented within a tool which operates as a standalone application. Through this tool, the method has been applied in three study cases. The findings obtained indicate promising potential of the method here introduced.

Based on Alves, A., Gersonius, B., Sanchez, A., Vojinovic, Z., and Kapelan, Z. (2018). "Multi-criteria Approach for Selection of Green and Grey Infrastructure to Reduce Flood Risk and Increase Co-benefits." Water Resources Management, Water Resources Management, 32(7), 2505–2522.

2.1 INTRODUCTION

Expected changes in future climate conditions include higher rainfall intensities in numerous places around the word (IPCC, 2012). Urban areas are among the most vulnerable regions to these changes, mostly because already more than half of world population lives in cities, and this number is projected to grow to 66% by 2025 (United Nations, 2014). A direct consequence of this growth is the increment of impervious surfaces, and therefore the intensification of runoff and heat stress. The excess of runoff brings about urban flooding and pollution of receiving water bodies, increasing the challenges in cities towards sustainable development (Jha *et al.*, 2012).

Appropriate flood risk management is crucial to mitigate flooding in urban areas. Nevertheless, decision-making to choose adequate strategies is a difficult process since it involves complex interactions between natural, social and built environments. The complexity of urban drainage systems, the quantity of available measures, the significant investments needed to implement these strategies, and the uncertainty about future conditions, increase decision-making complexity (Jha *et al.*, 2012; Simonovic, 2012).

In addition, decision makers are expected to consider criteria beyond the reduction of runoff when choosing these strategies. For instance, flood reduction strategies based on green infrastructure offer different co-benefits, which help to improve other adverse conditions besides flood risk, for instance reducing heat stress or water scarcity (CIRIA, 2013). The consideration of multiple benefits is an important element when planning sustainable systems, since it can help cities to be more resilient to changing future conditions (Lundy and Wade, 2011; IPCC, 2012).

Traditionally, the selection of measures to reduce flood risk is based on economic efficiency and suitability for local conditions, focusing mainly on traditional grey infrastructure (Vojinovic *et al.*, 2016a). Nowadays, it is becoming a well-accepted fact that this process needs to incorporate other elements, such as socio-ecological sustainability and adaptability to environmental changes. In other words, decision-making for flood risk management should be made from a holistic point of view, taking into consideration also green-blue measures and different aspects of urban environments: social, natural and built (Vojinovic, 2015). Therefore, it is needed to develop new methodologies and tools enabling better selection of flood mitigation measures.

In this chapter, a new method for selection of structural measures for flood risk reduction is introduced. The method takes into consideration green-blue and grey measures that can cope with different types of floods. The main objective is to support decisionmaking processes by allowing the selection of adequate measures in accordance to local conditions and preferences. The new methodology was implemented within a tool which operates as a standalone application. The methodology and the tool have been applied in three study cases within the PEARL (Preparing for Extreme And Rare events in coastal regions) EC-funded FP7 project (http://www.pearl-fp7.eu/).

2.2 SELECTION OF GREEN AND GREY MEASURES

New approaches in flood risk management are moving from centralized strategies that make use of built infrastructure, such as underground pipes; to multi-functional and distributed measures, which contribute to increase ecosystems resilience and restore the water cycle, for instance bio-retentions (Commission of the European Communities, 2009; Young *et al.*, 2011). The former approach is referred in this paper as grey infrastructure, while the later one as green infrastructure. Synonyms found in the literature to refer to the here called green infrastructure are: best management practices (BMP), low impact developments (LID), sustainable drainage systems (SuDS) and water sensitive urban design (WSUD) (Fletcher *et al.*, 2014).

Grey measures have the strengths of being reliable to cope with moderate rainfall events and are largely tested systems. Also, these measures offer opportunities such as the availability of enough methods for design and high acceptability. However, grey measures have the weakness of being single oriented towards flood management and provide low adaptability to future changes (CIRIA, 2013). In contrast, green measures use natural processes to cope with runoff excess, and at the same time offer multiple-benefits and improve adaptability (Tzoulas *et al.*, 2007; CIRIA, 2013). Still, whereas some above-ground green measures are effective to cope with extreme events (Fratini *et al.*, 2012; Recanatesi *et al.*, 2017), other options such as infiltration-based ones, are less reliable in handling medium and high return period rainfall events. Besides, green infrastructures applicability depends strongly on local characteristics, decision makers prefer a more traditional approach (Martin *et al.*, 2007; Naumann *et al.*, 2011; Chow *et al.*, 2014; Moura *et al.*, 2016) and tools for designing are still under development (Elliott and Trowsdale, 2007; Liu *et al.*, 2015).

Consequently, an approach combining green and grey infrastructure seems promising. By combining these two approaches the reliability and acceptability of grey systems can be mixed with multi-functionality and adaptability from the green side. Moreover, this arrangement could be very useful when retrofitting existent grey systems. The benefits of combining different measures has been already suggested by several authors (USEPA, 2000; Casal-Campos *et al.*, 2015; Voskamp and Van de Ven, 2015; Alves *et al.*, 2016a). However, a direct consequence is the increasing complexity of measures selection processes, given the numerous options, criteria and probable combinations.

Several methods and tools to help the selection of measures for flood risk management have been reported on the literature. Table 2.1 lists some of these works, detailing the type of method/tool and capabilities, and providing a source/reference where further information is available. Most of the methods and tools presented comprise rather simple and practical analysis to assist measures selection.

Table 2.1. Methods and tools for measures selection

Method/Tool	Type	Functions/Capabilities	Link/Reference
USDC Manual	Manual/ Guide	Decision trees and performance matrices for BMPs selection.	http://udfcd.org/volume-three
SuDS Manual	Manual/ Guide	Selection matrices based on qualitative assessment.	http://www.ciria.org/Resources/Free_ Publications/SuDS_Manual_C753.aspx
PEARL KB	Web application	Measures screening according to flood type, measure type, spatial scale and land use.	http://pearl-kb.hydro.ntua.gr/filter/ (Karavokiros et al. 2016)
Climate App	Web application	Measures screening according to problem to solve, land use, soil type and slope, scale and development stage.	http://www.climateapp.nl/
Urban green-blue grids	Web application	Measures screening according to benefit pursued.	http://www.urbangreenbluegrids.com /measures/
BMPSELECT	Software program	Multi-criteria ranking based on local suitability, runoff quantity and quality, additional benefits and cost.	(Jia et al. 2013)
BMP Selector	Software-aided approach	Analytical hierarchical ranking, considers site constraints, water quantity and quality, aesthetics and cost.	(Young et al., 2009; Young et al., 2011)
Multi-criteria decision aid	Method	BMPs multi-criteria ranking, considers cost, reliability, amenity, sustainability and water quality.	(Martin et al., 2007)
Multi-criteria decision framework	Method	Comparison of SuDS considering water quantity and quality, environmental benefits, energy saving and costs.	(Chow et al., 2014)
BMP DSS	Desktop application	Helps to evaluate alternatives, identifying cost efficient BMPs combinations for water quantity and quality.	(Cheng et al., 2009)
Adaptation support tool	Touch screen application	Selection of effective sets of blue-green measures based on water quantity and quality, cooling effect and cost.	(Voskamp and Van de Ven, 2015)

Among the methods presented in Table 2.1, the cases comprising green and grey solutions perform basic screening assessment (PEARL Project, 2015; Karavokiros *et al.*, 2016). While the methods developing rankings and comparative analysis only consider green infrastructure (Martin *et al.*, 2007; Cheng *et al.*, 2009; Young *et al.*, 2009, 2011; Jia *et al.*, 2013; Chow *et al.*, 2014). Five methods (Urban green-blue grids, BMPS Select, BMP Selector, Multi-criteria decision aid, and Multi-criteria decision support framework) consider a broad range of co-benefits when comparing measures, but none of them includes grey options. Only two methods (PEARL KB and Climate App) distinguish among different flood types.

Comparing with methods that generate measures rankings (BMPSELECT, BMP Selector and Multi-criteria decision aid), the method presented here has the advantage of considering grey measures besides green measures. Furthermore, in this case different types of floods are considered, while in the previous cases only pluvial floods were included. Finally, this method comprises a broader range of co-benefits and the possibility of decision makers to define preferences among these benefits.

The method presented here includes several steps which are described in detail in the next section.

2.3 METHODOLOGY

Multi-criteria Decision Analysis (MCDA) methods allow to structure complex problems and help a better understanding of the trade-offs implied. This type of analysis are helpful when decision-making situations with multiple and conflicting criteria arise. In particular, water management is characteristically a problem with multiple objectives, which makes MCDA an adequate and growingly used tool in these cases (Bana e Costa *et al.*, 2004; Hajkowicz and Collins, 2007). Results from these models should not be interpreted as the final solution to problems, but rather as information to understand the consequences of selecting a certain option. With this information, the decision-maker should be able to select suitable options from a set of available alternatives (Riabacke *et al.*, 2012).

The most relevant factors when choosing a MCDA method are consistence, transparency and simplicity. Hence, the method should give reliable results, following an understandable process by the user, and it should be easy to use. Among MCDA methods, multi-attribute utility theory is one of the most widely used in practical applications. This method displays the consequences of options selection according to predefined criteria. Besides, user preferences are considered through weights (Department for Communities and Local Government, 2009; Riabacke *et al.*, 2012). Another MCDA used for comparison and ranking of options allowing users judgments is the Analytical Hierarchy Process (Young *et al.*, 2009, 2011). However, this method offers low transparency, which is not recommended for decision-making processes involving many stakeholders (RPA, 2004).

Weighted summation is used in this work, which is considered as the simplest form of multi-attribute utility analysis (RPA, 2004). It is a linear method which gives relative differences among options by multiplying weights and scores; and adding up the resulting values. This method requires quantitative data and gives performance scores or rankings as result, with high transparency, simple computation and low cost. The difficulty of this method arises from selecting adequate scores to represent option's performance and to define weights (RPA, 2004). In this work, the scores are pre-defined based on an extensive literature review, while weightsare chosen by the user to represent local preferences.

The application of weighted summation methods contains the following steps: selection of appropriate criteria for options evaluation; definition of relative importance of each criterion (weighting); assessment of each option separately according to each criterion (scoring); combination of weights and performances to define the overall score for each option (ranking).

The criteria for evaluation of measures performance, as well as the scores to evaluate performance, are pre-selected in this case. This means, the user does not define these criteria neither the scores. The user answers questions about flood characteristics and local conditions, which are inputs for measures elimination or screening. Besides, the user chooses weights to establish which co-benefits are preferred in the area.

Figure 2.1 presents the different steps in this methodology that the decision maker needs to follow to develop the ranking of measures. Next, each of these steps is described.

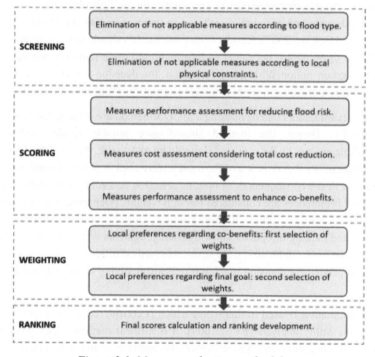

Figure 2.1. Measures selection methodology

2.3.1 Applicable Options Identification: Screening

The first step is the elimination of non-applicable options: screening. This step is important because it allows to focus the analysis only on feasible measures. The screening is based on flood type and local constrains. While all types of floods are included into this selection method, not all measures are suitable to reduce all types of floods (e.g. green roofs are not appropriate solutions to cope with fluvial floods). An important characteristic of this method is that more than one flood type can be considered simultaneously.

Regarding local constraints, several measures have implementation restrictions. As an example, if the site offers soils with very low permeability, measures based on infiltration are eliminated from the list of options. A detailed list of the criteria considered for screening is shown in Fig. 2.2-a.

2.3.2 Criteria for Performance Assessment

The criteria described here are used for scoring of measures based on performance assessment. These criteria should be easily evaluable, besides the number of criteria should be as low as possible but sufficient to make well-founded decisions. An important condition when using weighted summation is the independence of criteria. This means that the performance of an option regarding one criterion, can be defined without knowing its performance on other criteria (Department for Communities and Local Government, 2009).

In this case, several criteria are considered for the assessment of measures. The division of criteria into sub-groups is considered useful to clarify the process and to make the estimation of weights easier. Figure 2.2-b presents the division of criteria and sub-criteria in clusters through a value tree, which is used to show the hierarchy of criteria.

The criteria are clustered in three groups: functional goals regarding flood risk reduction, cost minimization and co-benefits enhancement. The last two criteria are further divided in two more levels of sub-criteria. This levels division could generate that criteria with more levels of division outscore the ones with less levels. One way to avoid this, and the most common method to combine scores and weights, is to work with simple weighted average scores (Department for Communities and Local Government, 2009). In this work, the score for each criterion is calculated as the average scores of its sub-criteria.

Regarding the sub-criteria considered, for flood reduction the reliability of each measure in front of rain events with different levels of return periods is assessed. Concerning cost, the sub-criteria are divided in three groups. Firstly, the costs that could be avoided choosing right measures according to local urban characteristics, for instance decentralized measures are preferred where large spaces are not available. Secondly, investment and maintenance costs, and finally the quantity of land required to implement the measure. Finally, twelve criteria are considered to evaluate co-benefits. These criteria are divided in five sub-groups: water quality, environmental benefits, livability enhancement, economic benefits and socio-cultural development.

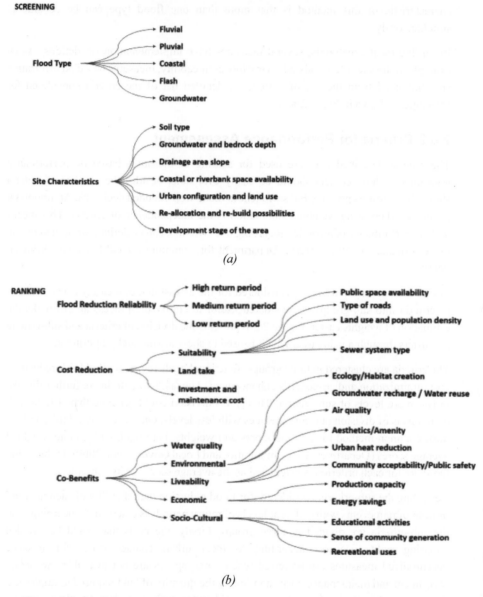

(a)

(b)

Figure 2.2. (a) Criteria considered for measures screening, (b) Value tree showing performance criteria for measures ranking

2.3.3 Performance Assessment: Scoring

Weighted summation is less appropriate to work with qualitative data, nevertheless in practice this issue can be solved assigning quantitative scores to qualitative information (RPA, 2004). In this work, a qualitative assessment of the effects of different options on each criterion was developed collecting data from literature review. Sources accessed include: USEPA (2000); Woods-Ballard *et al.* (2007); Center for Neighborhood Technology (2010); UDFCD (2010); CIRIA (2013). From this analysis, each option is assigned with low, medium or high impact level to each performance criteria.

For instance, USEPA (2000) affirms that bio-retention systems are less cost intensive than traditional conveyance structures. While Center for Neighborhood Technology (2010) states that pervious pavements have good performance improving water quality, and low performance on increasing recreational opportunities. Woods-Ballard *et al.* (2007) declares that infiltration trenches have medium performance on peak flow reduction, while are suitable for high-density residential areas. Most of the information collected from desk study refers to green infrastructure performance. Performance about grey measures is collected from literature (when available) and from experts' judgment.

To apply the weighted summation method, it is necessary to move from the obtained qualitative assessment to a quantitative description that reflects the proportional differences of impacts between options. This quantification was done assigning simple score levels based on the qualitative descriptors. Low score (chosen here as 1) represents very poor performance, this is for instance no water quality improvement, or high cost. Medium score (defined in this work as 3) represents the medium level of performance, and high score (represented by 5) is used in those cases where the measure has good performance. Using this data, performance matrices were built, with each row describing a measure and each column representing one criterion.

As mentioned before, in this method the scores are pre-defined, the user does not choose scores. This procedure is believed as simpler for the user, and is in line with the objective of developing a quick and user-friendly method for measures selection. Nevertheless, there is the possibility for the user to enter the performance matrix and overwrite performance scores if this is preferred.

2.3.4 Local Preferences Definition: Weighting

The relative importance of criteria is an essential concept in MCDA. At the same time, this is a stage in which decision-makers are intrinsically unsure, probably due to the difficulty of expressing preferences as numerical values. For this reason, the selection of weighting methods is crucial. Ratio weight procedures use values for representation of preferences. Simple elicitation methods in this category are Direct Rating and Point Allocation (Riabacke *et al.*, 2012). In the former, each attribute is rated from 0 to 100, whereas in in the last one a total of 100 points is distributed among the criteria. In this work, Direct Rating is used, which is believed to give more reliable weights since the result is not influenced by the additional cognitive effort needed to recall the remaining points to allocate (Riabacke *et al.*, 2012).

In the method here presented, the user defines weights in two different occasions. Firstly, to choose the importance of each co-benefit, which are divided in five sub-groups as was presented in Section 2.3.2. The user chooses scores from 1 to 10 for each sub-group of cobenefits, where 1 represents very low importance and 10 is used to represent the most preferred. In this way, the user follows a simple procedure to describe local preferences, then the values are normalized for further calculations.

As a final step, the user defines local priorities choosing weights for the three main goals considered: flood reduction, cost minimization and co-benefits development. The user chooses levels from 1 to 3 to describe the local importance of each objective. Again, higher values represent more important objectives.

2.3.5 Ranking

As last step, the overall score of each measure is calculated as the weighted average of scores of all criteria considered. Each general score calculation is the result of multiplying the screening analysis result, by weights times average of scores for each main goal. Meanwhile, the average score of co-benefits is calculated as the summation of the five co-benefit weights times the corresponding average scores. Next equations present these calculations.

Co-benefits weighted average score calculation is as follows:

$$S_{Co\_Be} = W_{WQ} * S_{WQ} + W_{En} * S_{En} + W_{Li} * S_{Li} + W_{Ec} * S_{Ec} + W_{S\_C} * S_{S\_C} \qquad (2.1)$$

where, SCo_Be is the average score of co-benefits; W_{WQ} is the selected weight and S_{WQ} is the average score in the case of water quality benefits; W_{En} and S_{WQ} are weight and average score for environmental benefits (which considers habitat creation, for instance to enhance biodiversity; groundwater recharge, achieved applying infiltration measures; rainwater reuse; and air quality improvement); W_{Li} and S_{Li} are weight and average score for livability enhancement (including amenity and aesthetic enhancement; heat stress reduction; community acceptability; and safety issues, for instance considering mosquitoes related to measures with runoff retention); W_{Ec} is the selected weight for economic benefits and S_{Ec} is its average score (considering production capacity, such as urban farms; and energy savings for example due to thermal isolation gained using green roofs); finally, W_{S_C} and S_{S_C} are weight and average score for socio cultural benefits (taking into account recreational activities;

educational uses, for example creating areas with native vegetation close to schools; and improvement of sense of community by developing leisure open areas within neighborhoods).

Overall score calculation for each measure is as follows:

$$Final\ Score = E_{Fl} * E_{LC} * \left(W_{Fl} * S_{Fl} + W_{Co} * S_{Co} + W_{Co\_Be} * S_{Co\_Be}\right) \qquad (2.1)$$

where, E_{Fl} and E_{LC} are the results of measures elimination according to flood type and local constraints respectively (these factors adopt values of 0 or 1, representing nonapplicable and applicable); W_{Fl} and S_{Fl} are the values of weight and average score regarding measures flood reliability; W_{Co} and S_{Co} are weight and average score in the case of cost reduction goal; while W_{Co_Be} and S_{Co_Be} are weight and average score regarding the achievement of co-benefits.

2.3.6 Sensitivity Analysis

A sensitivity analysis was performed to evaluate the impact of the weights chosen by users on measures selection results. Identifying the most influencing inputs, it is possible to focus on their definition. This allows to simplify the elicitation phase, reducing decision maker's effort by decreasing the cognitive load (Riabacke *et al.*, 2012).

The weights were used as criteria for the sensibility analysis in order to evaluate the influence of local preferences over final results. Besides, weights were chosen since they are a more frequent source of disagreement than scores (Department for Communities and Local Government, 2009). Monte Carlo method was applied to perform the sensitivity analysis, changing weight values and assessing the variation of the scores in the ranking of measures.

2.3.7 Implementation Tool

With the objective of facilitating the implementation of the described methodology, a software was coded in Delphi to develop a desktop application. The aim of this tool is to enable local stakeholders to follow the different steps of measures selection process and obtain the final ranking in a simple way. The procedure was developed to work in a catchment or neighborhood scale. More than 50 structural flood management measures are included in this tool and about 25 criteria are considered for screening and ranking. The structural measures included in the tool were defined by reviewing two existent screening applications: PEARL Knowledge Base (http://pearl-kb.hydro.ntua.gr) and Climate App (http://www.climateapp.nl/).

Users interact with the tool in two occasions. First, answering questions about flood type and local characteristics (see Appendix A). For these questions, answers are given as multiple choice. Secondly, the user chooses weights in two different stages. In the first stage, the user defines five weights in a range from 1 to 10 to describe local preferences regarding co-benefits. The upper part of Table 2.2 shows the five co-benefits considered and its descriptions as are given to the user in the tool.

Afterward, the user establishes weights to define the most important goals for the specific case. In this step three weights need to be selected in ranges from 1 to 3, representing low to high importance respectively. The lower part of Table 2.2 shows the explanations given in the tool to help the selection of these weights.

Table 2.2. Description of weights to be selected using the tool

WEIGHTING 1: LOCAL CO-BENEFITS PREFERENCES
(Please select the weight from 1 to 10 of each criterion according local preferences)

WATER QUALITY IMPROVEMENT: related to the general capacity of measures to remove runoff pollutants.

ENVIRONMENTAL BENEFITS: is the capacity of different measures to contribute with ecological diversity, species habitat creation, groundwater recharge, water reuse and air quality.

LIVEABILITY ENHANCEMENT: is the capacity of measures to improve local ahestetics, amenity and reduce urban heat island effect, while having community acceptability and low public safety risk.

ECONOMIC BENEFITS: related with the capacity of measures to allow production of food or materials, and the possibility of generate energy savings.

SOCIO-CULTURAL BENEFITS: is the capacity of measures to create educational spaces, generate community engagement and spaces for recreation.

WEIGHTING 2: MAIN GOALS IMPORTANCE

FLOOD REDUCTION: How frequent are flood problems affecting buildings and generating damages?
(Please,choose a value from 1 to 3 according to the importance of flood events in the area)

If flood events causing important damages occur once or more times per year, choose:	3
If flood events causing important damages occur once each 2 to 5 years, choose:	2
If flood events causing important damages are rare and occur less than once each 5 years, choose:	1

COST REDUCTION: Are there budget limitations significant enough as to give up co-benefits?
(Please, choose a value from 1 to 3 according to budget limitations for the project)

If there is reduced budget to implement measures and cost reduction is a main concern, choose:	3
If there are some budget restrictions but cost reduction is not the main concern, choose:	2
If there is budget availability to implement measures and cost reduction is not a concern, choose:	1

CO-BENEFITS: How relevant is to achieve benefits for the urban space besides flood reduction?
(Please, choose a value from 1 to 3 according to the importance of achieving co-benefits for the area)

If the achievement of co-benefits is also a main objective for this case, choose:	3
If co-benefits are not a main target but measures providing them are still preferred, choose:	2
If the achievement of co-benefits is not important in this case, choose:	1

2.4 RESULTS AND DISCUSSION

2.4.1 Study Areas

The presented method has been applied in three case studies: Marbella, a coastal city in south Spain; Ayutthaya, located 80km north of Bangkok in Thailand; and Sukhumvit, a business area located in eastern Bangkok. The interest behind analyzing three cases lies in the possibility of testing the method for three different flood types. Moreover, there are other differences among these cases, while Sukhumvit is a highly urbanized and densely populated business area, Marbella is a more residential also highly urbanized area, and Ayutthaya is a sub-urban and less populated area. Therefore, the three cases represent different examples of urban spaces with diverse problems to solve.

In the examples presented here, decision makers followed the method answering questions about local characteristics and selecting weights. Firstly, the tool was presented and explained during group meetings with local stakeholders (see Figure 2.3). Then the process was followed individually, answering the questions through an online questionnaire. Afterwards, the answers were processed to obtain consensus among different stakeholders.

Figure 2.3. Group meetings to implement the tool with stakeholders

Local characteristics obtained for the three case studies are presented in Table 2.3. These characteristics are divided in: data required for measures screening and data necessary for measures ranking.

Table 2.3 also presents the weights chosen in each case. In general, it can be observed that flood reduction is selected as the most important goal, while cost decrease is principally selected as second goal. Only in the case of Sukhumvit the enhancement of co-benefits is selected as a relevant goal. It is not possible to identify general preferences among the different co-benefits.

Table 2.3. Characteristics of study areas and preferences weights (inputs to the tool)

Case Study:	Marbella	Ayutthaya	Sukhumvit
Characteristics for screening:			
Flood type	Flash	Fluvial	Pluvial
Soil infiltration capacity	Medium	High	High
Groundwater depth	More than 1m	More than 1m	More than 1m
Bedrock depth	More than 1m	More than 1m	More than 1m
Draiange area slope	Lower than 5%	Lower than 5%	Lower than 5%
Available coastline/riverbank space	NA	Mainly urbanized	NA
Urban Configuration	Highly urbanized	Low urbanized	Highly urbanized
Realocation or re-build on pilars	Not possible	Not possible	Not possible
Development of area under risk	Developed	Developed	Developed
Characteristics for ranking:			
Public space availability	Less than 25%	More than 25%	Less than 25%
Availability of linear spaces in roads	No availability	Availability	No availability
Average population density	> or = 100p/ha	< 100p/ha	> or = 100p/ha
Main land use	Residential	Residential/Touristic	Residential/Commercial
Type of sewer system	Combined	Combined	Combined
WWTP/CCO volume reduction	Yes	No	No
Co-benefits preferences weights:			
Water quality improvement	6	7	8
Environmental benefits	3	6	8
Liveability enhancement	4	7	9
Economic benefits	2	7	9
Socio-Cultural development	3	7	9
Goals weights:			
Flood reduction	3	3	3
Cost minimisation	2	2	2
Co-benefits improvement	1	1	3

2.4.2 Tool Implementation Outcome

The screening of measures for each case study was performed according to local characteristics (Table 2.3). Results are presented in Table 2.4, which depicts for each case study, if the measures are eliminated from the list of applicable measures or remain for further analysis.

Table 2.4. Remaining (grey) and eliminated (white) measures after screening process, for each study area (M:Marbella, A: Ayutthaya, S: Sukhumvit)

Measure	Description	Area		
		M	A	S
Closed conduits	Traditional underground pipes, tubes and tunnels	■		■
Demountable barriers	Temporal demountable flood protection, for instance sandbag wall	■	■	
Underground Storages	Subsurface structures to capture and storage runoff temporarily	■		■
Dry Flood Proofing	Sealing buildings to prevent the entrance of stormwater	■	■	■
Bypasses	Provides extra discharge capacity for rivers and canals	■	■	
Hollow roads	Increment of roads transport capacity by hollow roads or rised curbs	■		■
Wet Flood Proofing	Allows stormwater entering buildings with waterproof interior		■	■
Dikes rising	Elevation of existent dikes to improve low-lying areas protection.		■	■
Open detention basin	Surface storage basins that provides flow control		■	■
Polder	Low-lying area enclosed by dikes, pumpling is often needed		■	
Floating constructions	Buildings resting on the ground which can float during flood events		■	■
Non-return valves	Valves installed in pipes vulnerable to backflow in flood conditions		■	■
Pumping systems	Applied to remove exces of runoff when there is not natural flow	■	■	■
Rain gardens/Bio-retention	Vegetated depressions for runoff infiltration/evapotranspiration			■
Green Roofs	Roofs covered with vegetation to intecept precipitation			■
Rainwater disconnection	Runoff is transported through surface or infiltrated			■
Water/Blue Roofs	Temporary storage of rainwater in flat roofs			■
Rainwater Harvesting	Rainwater from roofs and hard surfaces is stored allowing reuse			■
Infiltration pipes	Underground storage and infiltration, as well as transport system			■
Green Walls/Facades	Vertical vegetation areas and gardens			■
Pervious Pavements	Pavement that allows rainwater infiltration and temporary storage			■
Soakaways/Inf. boxes	Filled exavations for runoff attenuation and infiltration			■
Paved surfaces reduction	Increment of green areas to improve infiltration and reduce runoff			■
Infiltration Trench	Rock-filled trench with no outlet for storage and infiltration of runoff			■
Infiltration Trench (underdrain)	Infiltration trench with conveyance pipe for low perviousness soil			■
Infiltration basin	Vegetated depressions for runoff storage and gradual infiltration			■
Infiltration basin (underdrain)	Infiltration basin with pipe system for cases of low perviousness soil			■
Open Gutters	Non-permeable open drain located along a roads			■

The rankings of applicable measures and results of sensitivity analysis for each case study are shown in Figure 2.4. Green measures are represented using green bars, while grey bars represent grey measures. The ranges of scores go from zero to one hundred.

Sensitivity analysis results are showed as black lines in Figure 2.4. These lines cover the ranges in which scores move when weights are changed. This ranges are the result of applying the Monte Carlo analysis. The analysis was performed varying at the same time the eight weights inside their limits (1 to 10 for preferences and 1 to 3 for goals), and assessing the changes in final scores through the use of Equations 2.1 and 2.2. For this, the model was run 5000 times with different combinations of weights.

Observing the rankings obtained, two main characteristics of the method stand out. First, centralized options, such as detention basins, are preferred in cases where population density is low and public spaces are available (Ayutthaya); while decentralized options, such as rain gardens or green roofs, are chosen when space availability appears as a main constraint (Sukhumvit). This result confirms that the method successfully considers local space availability and selects measures accordingly.

Second, both green and grey options appear within the higher rankings obtained, suggesting that combinations of these two approaches could lead to efficient and sustainable solutions. This result is a consequence of the diversity of factors included into the selection process. While some factors promote the selection of gray measures, such as the reliability in front of different return periods and consideration of local constraints; other factors benefit green measures selection, as co-benefits enhancement.

Regarding sensitivity analysis results, overall scores fluctuate between 25% and 100% of total maximum score when weights are changed, depending on the measure and case. Consequently, the selection of weights has a significant impact on the ranking results. This emphasizes the influence of the choices made by the user on the result. A direct consequence of this result is the importance of providing the user with the necessary elements to allow a simple and informed selection of weights. Hence, the importance of using a simple weighting method (Direct Rating), and of providing to the user clear information about the meaning of each weight (Table 2.2).

To better understand the impact of weights on final scores, it is relevant to analyse the influence of each weight on final results. Figure 2.5 shows this influence for the measures previously introduced. It can be observed that the three goals weights together are responsible for over 99% variance in scores for most of grey measures. The fourth weight, called cobenefits preferences in Figure 2.5, represents the addition of the five weights chosen to define preferences among co-benefits.

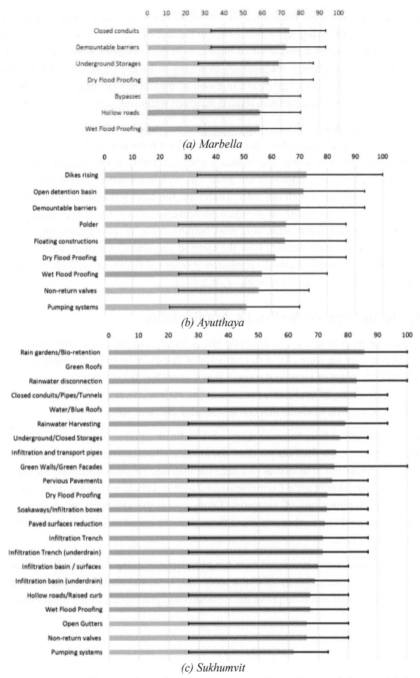

(a) Marbella

(b) Ayutthaya

(c) Sukhumvit

Figure 2.4. Ranking results and range of scores variance from weights sensitivity analysis (grey infrastructures: grey bars, green infrastructures: green bars)

The impact of co-benefits preferences is low in all cases, although in the case of green measures this value has higher influence than in the case of grey measures. This is because co-benefits preference weights are used implicitly in the calculation of overall scores (see Equation 2.2). Consequently, the preferences among co-benefits have very little influence for the final ranking, only significant in cases of measures that offer high co-benefits impact, for instance paved surfaces reduction, green walls and infiltration trenches.

The most influencing goal weight on overall score variance differs for each measure (Figure 2.5). Generally, grey measures will have flood reduction as the most influencing weight, while green measures experience higher variability from changing the weight associated to co-benefits improvement goal. The weight related with cost minimization affects mainly measures with low cost. Therefore, the most influencing weight for each measure will correspond with its strongest performance, this is the higher average score per category, either flood reduction, cost minimization or co-benefits improvement performance. This result reflects how the method calculates overall scores. Moreover, these outcomes prove the method's consistency since the final score is highly dependent on the capacity of the measure to achieve each goal.

2.4.3 Reflection on Method's Applicability

The main strength of this method is reflected in its capacity to analyze the performance of green and grey measures for different types of floods. Moreover, the analysis is done from a holistic point of view in the sense that it considers flood reduction, costs and co-benefits for ranking of applicable measures. The review of current literature shows that this is one of the first methods that combines these characteristics.

The results obtained from the case study work indicate a promising potential of the new method. However, the final output does not provide a unique and defined strategy (seen here as a combination of measures) for decision making. This was done with the idea that the final strategy should arise from discussions with stakeholders and the deeper analysis of preferred combinations of measures. Furthermore, the efficiency of these possible combinations could be further evaluated through more complex analysis before arriving with a final decision.

Even though the results presented here are promising, a more elaborate discussion and validation of the methodology (and the tool) with local stakeholders is yet to be done. Such validation will aim to improve the workings and practicality of the methodology.

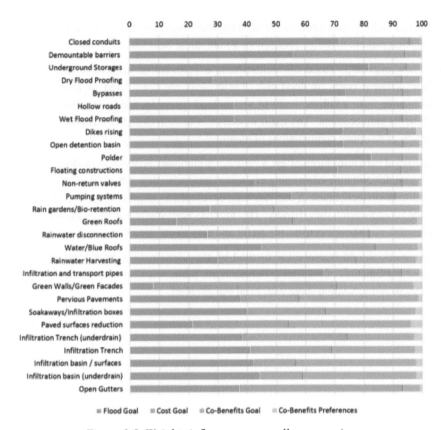

Figure 2.5. Weights influence on overall score variance

2.5 CONCLUSIONS

Although the advantages of including green infrastructure as part of flood management plans have been widely discussed and acknowledged, often decision makers are still reluctant to select this approach confidently. Moreover, recent discussions about sustainability of drainage systems focus on the use of green infrastructure, without considering the combination of green and grey options. However, this combination appears as the best way to ensure reliability in front of extreme events, and at the same time obtain co-benefits provided by green infrastructure.

A novel method for general assessment and selection of green and grey measures to reduce flood risk was presented in this chapter. The method presented here has several advantages comparing with previously developed methods. Firstly, it considers grey measures besides green measures. Secondly, it includes different types of floods into the analysis. Finally, this method comprises a broader range of co-benefits, and the possibility of decision makers to define preferences among these benefits.

This new method is based on multi-criteria assessment of infrastructures oriented to reduce different types of flood risk. Three goals are considered to assess measures performance from a holistic point of view: flood risk reduction, cost minimization and co-benefits enhancement. The main objective is to help decision makers, assisting the selection of adequate combinations of measures.

The method here introduced offers a valuable preliminary analysis to facilitate the use of more complex and time-consuming evaluation methods, such as hydrodynamic modelling or optimisation techniques. A tool has been developed to simplify the use of this method, making its application user friendly. This tool has been applied on three study cases. The results obtained show a consistent, useful and easy to use method.

Future work should improve the method by the addition of a module to analyse and suggest combination of measures. Furthermore, future work should also include a final discussion and validation stage with local stakeholders.

3

CONSIDERING STAKEHOLDERS PERCEPTIONS FOR GREEN-BLUE INFRASTRUCTURE SELECTION

Traditional approaches for flood management offer options with low sustainability. As a response, the use of non-traditional drainage measures, also called green infrastructures, has been increasingly suggested in the last years. One important reason for their increasing popularity has been the co-benefits that they offer to the environment. The development of an effective planning for sustainable urban drainage systems is a complex process that needs the involvement of multiple stakeholders. Moreover, the measures to be adopted should be evaluated considering their potential to achieve multiple benefits related to human well-being, rather than just to flood risk management. This chapter provides a framework for the selection of green infrastructures on the basis of a co-benefits analysis. The aim is to include the achievement of co-benefits and human well-being into decision-making for flood management. To achieve this, co-benefits are analysed from an ecosystem services point of view. The focus of this method is to consider stakeholders' perceptions to define the most important benefits to be enhanced. The application of the framework presented here to a case study in Ayutthaya, Thailand, shows the importance of including different stakeholder's opinions. In addition, it shows that decision makers should consider locally defined co-benefits as well as flood risk reduction when defining which green infrastructures to apply.

Based on Alves, A., Patiño Gómez, J., Vojinovic, Z., Sanchez, A., and Weesakul, S. (2018). "Combining Co-Benefits and Stakeholders Perceptions into Green-blue Infrastructure Selection for Flood Risk Reduction." Environments, 5, 29; doi:10.3390/environments5020029.

3.1 INTRODUCTION

In the last decades, flood management systems have been under growing pressure because of the population growth and its associated impervious surface expansion. Moreover, as a result of climate change, a higher rainfall intensity is expected in many regions around the globe. The combined effects of these two drivers and their tendency suggest an important increase in future flood risk levels (Mynett and Vojinovic, 2009; IPCC, 2012). Traditional approaches for flood management offer options with low sustainability and flexibility, which are needed to cope with an uncertain future. As a response, the use of non-traditional drainage measures, also called green-blue infrastructure (GBI), has been increasingly suggested in the last years. One important reason for GI increasing popularity has been the co-benefits that they offer to the environment (CIRIA, 2013). These benefits include environmental and socioeconomic aspects, such as the reduction of energy and water consumption, biodiversity enhancement, and health benefits, among many others (Center for Neighborhood Technology, 2010; European Commission, 2012a; CIRIA, 2013).

The development of an efficient planning and design framework for sustainable urban drainage systems is a complex process. This process needs an interdisciplinary approach and the involvement of multiple stakeholders, who have often conflicting interests (Yazdanfar and Sharma, 2015). To develop effective strategies for future scenarios, it is important to identify the best measures to be applied in each context. To accomplish this, the measures should be evaluated considering their potential to achieve multiple benefits related to human well-being, rather than just to the management of stormwater (Vojinovic, 2015).

According to Lundy and Wade (2011), multifunctional landscapes designed for ecosystem service (ES) provision can help to be more sustainable and more resilient to the changing future conditions. Since ecosystem services, human well-being, and the achievement of co-benefits are intrinsically related, the pursuit of multiple benefits appears as an important element when planning sustainable stormwater management systems.

Several methodologies have been developed to help green infrastructure selection for stormwater management (Martin *et al.*, 2007; Cheng *et al.*, 2009; Young *et al.*, 2011; Chow *et al.*, 2013; Jia *et al.*, 2013). While these works have considered co-benefits when selecting measures, the definition of the main co-benefits is not based on stakeholders' preferences. Meerow and Newell (2016) performed a stakeholders' survey to prioritize benefits for different landscapes and locate where these benefits were needed. Schifman *et al.* (2017) developed a framework to integrate networks of organizations into GBI projects to achieve multiple benefits. This work is centred on the combination of inputs from organizations to reach collaborative decision-making. Other works compared different solutions using more complex analyses (Jia *et al.*, 2012; Alves *et al.*, 2016b). Considering co-benefits and stakeholders' perceptions appears to be effective in performing a preselection of measures before applying more complex methods, enhancing their effectiveness.

In this chapter a framework for the evaluation and selection of green infrastructures based on co-benefits analysis is proposed. The aim is to include the achievement of co-benefits and human well-being into the decision-making processes related to flood management, considering both local aspects and stakeholders' preferences when defining the most important benefits to be enhanced. The focus is on the identification of key benefits through stakeholders' perceptions analysis as a central aspect to select flood risk reduction strategies.

3.2 LINKING GREEN INFRASTRUCTURE, CO-BENEFITS, ECOSYSTEM SERVICES, AND HUMAN WELL-BEING

It has been proved that GBI are effective in reducing flood risk (Fratini *et al.*, 2012; Moura *et al.*, 2016; Recanatesi *et al.*, 2017). The traditional approach for GI selection is based on runoff reduction assessment, cost minimisation, and suitability. However, more and more decision makers are expected to consider other factors when choosing flood management strategies. For instance, GBI offer different co-benefits which contribute to cope with other problems of urban environments besides flood risk (CIRIA, 2013), helping to improve the quality of life of the citizens. Examples of these problems are air pollution and heat stress.

Moreover, GBI impact positively on the health of ecosystems through the provision of ecosystem functions and services, which offer the environmental conditions needed to improve human well-being. Different components of human well-being are physical, psychological, social, and community benefits (Tzoulas *et al.*, 2007). In summary, ES and co-benefits are provided by GBI promoting healthy environments, and contribute to environmental, social, and economic benefits to people in contact with them (Figure 3.1).

Ecosystem services are seen here as necessary services to achieve ecosystem health and human well-being, and they can be provided by natural or constructed environments (GBI in the latter case). Co-benefits are seen as the benefits that can be achieved by applying constructed GBI. Consequently, ES are seen here as a much broader and general concept than co-benefits. For instance, wetlands and green areas (natural or built) will contribute to the regulation and maintenance of biota and ecosystems mediation (Table 3.1). Meanwhile, the construction of urban wetlands and green spaces offer the co-benefits of air and water pollution abatement, besides flood risk reduction.

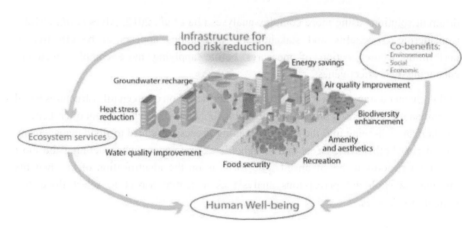

Figure 3.1. Links between GI for flood risk reduction, ecosystem services, human well-being, and co-benefits (Tzoulas et al., 2007; Haines-Young and Potschin, 2010)

3.2.1 Ecosystem Services and Human Well-Being

The enhancement of co-benefits is intrinsically related to the achievement of human well-being. Meanwhile, the achievement of human well-being is closely connected with the provision of ecosystem services (ES), which are defined as the benefits that people obtain from ecosystems (see Millennium Ecosystem Assessment, 2005; Defra, 2007; Haines-Young and Potschin, 2012).

One of the first frameworks for the evaluation of ES was proposed by de Groot *et al.* (2002), where ES were grouped in four categories: regulation, habitat, production, and information. This framework was later used by Millennium Ecosystem Assessment (2005) as the basis for its classification in provisioning, regulating, cultural, and supporting services. In this classification, supporting services are considered as necessary to produce all other services, having an indirect impact over people.

The Millennium Ecosystem Assessment framework for the assessment and integration of services, goods, and benefits produced by ecosystems continues to have a significant impact (Lundy and Wade, 2011). However, Haines-Young and Potschin (2012) applied the typology of ES suggested by Millennium Ecosystem Assessment framework to develop a more explicitly hierarchical structure. The authors considered the same three categories used by Millennium Ecosystem Assessment: provisioning, regulating, and cultural services, though supporting services are not considered to avoid double counting.

The importance of describing and classifying ES is founded on the close connection between this concept and human well-being, since the enhancement of ES is the basis to achieve human well-being. The components of human well-being have been classified in three groups: security, basic materials and production factors, and health and good social relations and health (Millennium Ecosystem Assessment, 2005; Staub *et al.*, 2011), (see Figure 3.2).

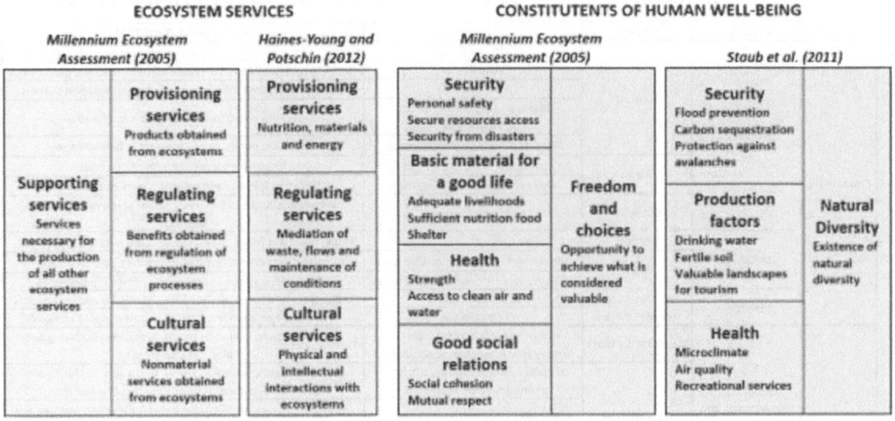

ECOSYSTEM SERVICES
Millennium Ecosystem Assessment (2005)
Haines-Young and Potschin (2012)
Supporting services
Services necessary for the production of all other ecosystem services
Provisioning services
Products obtained from ecosystems
Regulating services
Benefits obtained from regulation of ecosystem processes
Cultural services
Nonmaterial services obtained from ecosystems
Provisioning services
Nutrition, materials and energy
Regulating services
Mediation of waste, flows and maintenance of conditions
Cultural services
Physical and intellectual interactions with ecosystems
CONSTITUTENTS OF HUMAN WELL-BEING
Millennium Ecosystem Assessment (2005)
Staub et al. (2011)
Security
Personal safety
Secure resources access
Security from disasters
Basic material for a good life
Adequate livelihoods
Sufficient nutrition food
Shelter
Health
Strength
Access to clean air and water
Good social relations
Social cohesion
Mutual respect
Freedom and choices
Opportunity to achieve what is considered valuable
Security
Flood prevention
Carbon sequestration
Protection against avalanches
Production factors
Drinking water
Fertile soil
Valuable landscapes for tourism
Health
Microclimate
Air quality
Recreational services
Natural Diversity
Existence of natural diversity

Figure 3.2. Classification of ecosystem services and components of human well-being (Millennium Ecosystem Assessment, 2005; Haines-Young and Potschin, 2012; Staub et al., 2011)

Haines-Young and Potschin (2010) developed an approach linking ES, ecosystem functions, human well-being, and related benefits. In this scheme, ES are generated by ecosystem functions, which are based on biophysical structures (the services considered as supporting in the Millennium Ecosystem Assessment framework). Afterwards, the effect of ES on the sociocultural context is achieved through the impact on human well-being by the release of benefits and their associated values (see Figure 3.3).

Ecosystems and biodiversity
Biophysical structure
Ecosystem functions
Ecosystem services
Human well-being
Benefits
Value

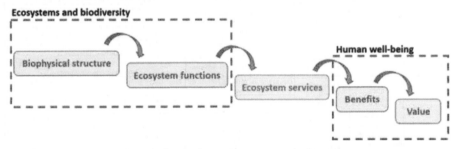

Figure 3.3. Relation between ecosystem functions, services, benefits, and value, adapted from Haines-Young and Potschin (2010)

Table 3.1. ES classification, based on Haines-Young and Potschin (2012), and examples of green infrastructure providing these services (Woods-Ballard et al., 2007; Center for Neighborhood Technology, 2010; UDFCD, 2010; CIRIA, 2013; Horton et al., 2016)

SECTION	DIVISION	GROUP	CLASS	GREEN INFRASTRUCTURE
Provisioning	Nutrition	Biomass	Cultivated crops	Public/Private green spaces, green walls, green roofs, urban trees
			Wild Plants	Green spaces, wetlands, rain gardens, bioswales and green walls
			Wild Animals	Green spaces, wetlands, retentions ponds and open chanels
			Plants and algae from in-situ aquaculture	Retention ponds, open channels, urban wetlands
			Animals from in-situ aquaculture	Retention ponds, open channels, urban wetlands
		Water	Surface water for drinking uses	Rainwater harvesting, retention ponds
			Groundwater for drinking uses	Infiltration surfaces and trenches, pervious pavements
		Nutrients	Nutrients for plants	Urban agriculture, urban trees, urban parks, rain gardens, wetlands
	Materials	Biomass	Fibres and other materials	Urban trees and forest, green spaces, wetlands and green roofs
		Water	Surface water for non-drinking uses	Rainwater harvesting, retention ponds, open channels
			Groundwater for non-drinking uses	Infiltration surfaces and trenches, pervious pavements
	Energy	Biomass-based energy sources	Plant-based sources	Urban trees and forest, green spaces, wetlands and green roofs
			WW-based sorces	Wetlands, vegetation fertilization in green spaces using WW
Regulating and maintenance	Mediation of waste, toxics and other nuisances	Mediation by biota	Bio-remediation, filtration, sequestration, storage, accumulation	Urban wetlands, retention ponds, bioswales, buffering and bioretention areas
		Mediation by ecosystems	Filtration/sequestration/storage/ accumulation by ecosystems	Urban wetlands, retention ponds, bioswales, buffering and bioretention areas, infiltration areas and trenches
			Enhancement of pollutants removal systems	Green spaces, rain gardens, bioswales, infiltration surfaces and trenches, retention ponds, wetlands, pervious pavements and buffering areas
			Mediation of smell, noise, visual impact	Green walls/fecades, green spaces, green noise barriers, urban trees
	Mediation of flows	Liquid flows	Hydrological cycle and water flow maintenance	Infiltration areas, trenches and pavements, rainwater disconnection, open channels
			Flood protection	All green measures allowing storage, infiltration, convey enhancement and imperviousness reduction in general
			Combined Sewer Overflow reduction	Green measures allowing storage, infiltration and runoff flow reduction in general
		Gaseous/air flows	Storm/wind protection	Green barriers, urban trees and forest
			Ventilation and transpiration	Urban parks and forest, green spaces, rain gardens, green roofs and walls
	Maintenance of physical, chemical, biological conditions	Lifecycle maintenance, habitat protection	Pollination and seed dispersal	Green spaces, green roofs and walls, wetlands, rain gardens
			Maintaining nursery populations and habitats	Green spaces, retention ponds, open channels, wetlands and buffering areas
		Pest and disease control	Pest control	Negative impact of wetlands, retention ponds, green sapces and buffering areas
			Disease control	Negative impact of wetlands, retention ponds and buffering areas
		Soil formation and composition	Weathering processes	Urban wetlands, bioswales, retention ponds
			Decomposition and fixing processes	Urban wetlands, bioswales, rain gardens, retention ponds, buffering areas
		Water conditions	Chemical condition of freshwaters	Infiltration surfaces and trenches, urban wetlands, pervious pavements, buffering and bioretention areas.
		Atmospheric composition and climate regulation	Global climate regulation by reduction of greenhouse gas concentrations	Green spaces, green roofs and walls, wetlands, rain gardens, urban trees, bioswales
			Micro and regional climate regulation	Green spaces, green roofs and walls, pervious pavements, retention ponds, open channels
Cultural	Physical and intellectual interactions with biota, ecosystems, and land-/seascapes	Physical and experiential interactions	Experiential use of plants, animals and land-/seascapes	Green spaces, green roofs and walls, wetlands, retention ponds, open channels
			Physical use of land-/seascapes	Green spaces, urban parks, retention and detention ponds, open channels
		Intellectual and representative interactions	Scientific	Green measures allowing monitoring of runoff flow and pollution reduction, as well as other co-benefits
			Educational	All visible green measures allow awareness and engagement through education opportunities
			Heritage, cultural	Opend detention basins with multifunctional uses, open channels and retention ponds
			Entertainment	Green spaces, urban parks, retention ponds, open channles, multifunctional detention basins
			Aesthetic	Green spaces, urban trees, ran gardens, retention ponds, open channels, wetlands
	Spiritual, symbolic interactions with biota and ecosystems	Spiritual and/or emblematic	Symbolic	Green spaces and parks, retention ponds
			Sacred and/or religious	Green spaces and water surfaces
		Other cultural outputs	Existence	Green spaces, water surfaces, infiltration and buffering measures
			Bequest	Green spaces, water surfaces, infiltration and buffering measures

3.2.2 Ecosystem Services and Green-Blue Infrastructure

According to Tzoulas et al. (2007) green infrastructure includes all natural, semi-natural and artificial networks of ecological systems at all spatial scales, which have important multifunctional and habitat interconnection roles, contributing to biological diversity conservation and maintaining integrity of habitats. Consequently, these infrastructures have an impact on human health and well-being by improving health of ecosystems. Different GBI affect different ecosystem services, it is important to understand the linkage between them to understand the impact that applying GBI would have on ES and well-being (see Table 3.1).

Conforming to Lovell and Taylor (2013), the development of multifunctional GBI which integrates the achievement of different ES, is essential to obtain resilient cities in front of future challenges, such as climate change. The implementation of multifunctional GBI helps the development of adaptive strategies to cope with unknown future conditions. Furthermore, it is crucial to reach a holistic urban planning process, involving different stakeholders' groups into the decision making stage.

3.2.3 Green infrastructure and co-benefits

Co-benefits are seen here as the benefits that can be obtained from applying GBI, besides flood management. Several co-benefits can be achieved through green infrastructure implementation. Next, we present some of these co-benefits and examples of green measures capable of deliver them.

Water quality of receiving bodies: Stormwater runoff carries pollutants to receiving waters. GI, such as bio-swales and pervious pavements use vegetation and soil to filtrate this runoff (UDFCD, 2010; Horton *et al.*, 2016).

Groundwater recharge: GI that allow infiltration provide groundwater recharge, which is important where groundwater levels are reduced due to over abstraction or dry conditions (Horton *et al.*, 2016).

Biodiversity and ecology enhancement: Vegetation provides habitat for many animals. Large-scale green infrastructure, such as parks and wetlands, help wildlife restoration (Woods-Ballard *et al.*, 2007; Center for Neighborhood Technology, 2010; Horton *et al.*, 2016).

Heat stress reduction: Green areas and water spots, such as lakes, rivers and fountains, moderate temperatures and help to mitigate urban heat stress effect (Center for Neighborhood Technology, 2010; WHO, 2016).

Air quality improvement: Trees and green areas produce oxygen and help to filter harmful air pollutants (Center for Neighborhood Technology, 2010; WHO, 2016).

Amenity and aesthetics: Vegetation, water and wildlife provide natural and beautiful environments where people feel more comfortable (Center for Neighborhood Technology, 2010; WHO, 2016).

Recreation: Green spaces offer open places for human recreation. Having access to green open spaces reduces health issues and improves well-being (WHO, 2016).

Health: WHO (2016) recommends at least 10 m^2 of green spaces per inhabitant in urban areas. Physical activity in natural environments helps to reduce mental health issues.

Food security: The creation of urban farming spaces in green areas and green roofs is considered a strategy to improve food security in cities (Center for Neighborhood Technology, 2010).

Rainwater harvesting: Collected water through rainwater barrels can be used for outdoor irrigation, reducing significantly the use of potable water (Horton *et al.*, 2016).

Pumping and treatment reduction: Reducing runoff, the amount of water to combined drainage networks is reduced, decreasing pumping and wastewater treatment costs (Horton *et al.*, 2016).

Saving energy in buildings: Some technologies like green roofs or vertical gardens on buildings work as temperature isolation layers, reducing the necessity of cooling and heating (Macmullan and Reich, 2007; Center for Neighborhood Technology, 2010).

Real estate value appreciation: Increasing vegetation and trees cover increase property values in the area, benefiting both developers and homeowners (de Groot *et al.*, 2002; Derkzen *et al.*, 2015).

3.3 METHODOLOGY

The objective of the framework presented here is the selection of GBI combinations according to local conditions and preferences expressed by stakeholders. This is achieved following several steps, some of them focus in local physical conditions, and other based on local preferences (Figure 3.4).

The first step consists on finding and analysing relevant information about local characteristics and current situation of the area under study. The output of this step is an inventory with the amount and type of potential sites for GBI measures application, reached through land use analysis. The main objective is to assess the feasibility of potential GBI applicability in the area according to local land use. This analysis is particularly relevant for areas that are starting to develop plans using GBI measures. Through this analysis, the local potential for GBI measures usage can be determined.

The second step is to identify the most important benefits to be enhanced in the study area according to local needs and stakeholders' preferences. The importance of this step lies on the complexity of decision making processes due to the variety of stakeholders' points of view. While there is no agreed method for estimating the benefits of GBI, it is recommended to perform this process from large scale to a site specific perspective (CIRIA, 2013; Jia *et al.*, 2015). In order to achieve this, firstly, attention is paid to local regulations and previous studies developed in the area under study. Secondly, the analysis is focused on local stakeholders to confirm or correct the information found on previous steps (Patiño Gómez, 2017).

The expected result is the identification of which measures are applicable and which are the main concerns in terms of social, environmental and economic aspects in the area. From this analysis, a list of preferred measures to enhance key co-benefits for each land use type is obtained. As final step, the capacity of different GBI measures to reduce flood risk is considered before defining the final strategy of GBI measures combination for the area under study.

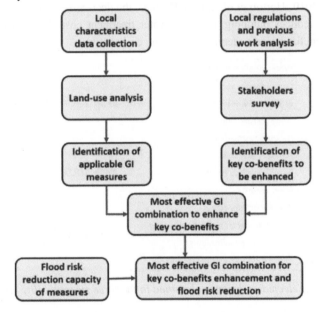

Figure 3.4. Methodological steps for selection of GBI combination

3.3.1 Co-benefits classification

With the objective of clearifing the analysis and to help the understanding of stakeholders, the co-benefits presented in Section 3.2.3 are divided in three groups for the analysis of perceptions. These groups represent three sustainability dimensions: environmental, social and economic benefits (see Figure 3.5).

Among environmental benefits of GBI are water quality improvement of runoff and receiving waters, infiltration to recharge aquifers, heat stress or temperature reduction, creation of better places for nature, enhancement of biodiversity and air quality improvement (Woods-Ballard *et al.*, 2007; WHO, 2016).

GBI measures improve life quality by making environments more vibrant, visually attractive and delivering recreation and education opportunities (Woods-Ballard *et al.*, 2007). Furthermore, urban parks and gardens provide sites for physical activity, social interaction and recreation (WHO, 2016). Consequently, amenity and aesthetics, recreation and health, and food security are considered as social benefits.

The use of green infrastructure brings economic benefits by storing water for reuse or reducing energy consumption for cooling or heating buildings, while aesthetics and amenity enhancement add value to surrounding buildings (Center for Neighborhood Technology, 2010; Horton *et al.*, 2016). Besides, the decrease of runoff flow going into pipe systems reduces pumping and wastewater treatment costs (Horton *et al.*, 2016). These four benefits are considered as economic co-benefits.

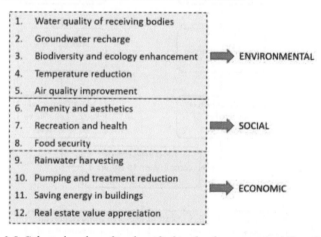

Figure 3.5. Selected co-benefits classified in the three sustainability dimensions

3.3.2 Establishing quantitative indicators to assess co-benefits

In this section, we present quantitative indicators representing the capacity of different GI to provide the co-benefit introduced in Section 3.2.3. These indicators were estimated using different sources of literature (Woods-Ballard *et al.*, 2007; Berghage *et al.*, 2009; Shoemaker *et al.*, 2009; Center for Neighborhood Technology, 2010; UDFCD, 2010; CIRIA, 2013; Jia *et al.*, 2013; DEFRA, 2016). For instance, pervious pavements have good performance for improving water quality and medium performance for groundwater recharge (Center for Neighborhood Technology, 2010). CIRIA (2007) affirms that infiltration trenches have medium to low amenity potentiality. Jia et al. (2013) declares that rain barrels have high performance in providing the benefit of rainwater reuse, while bio-retention areas are a good option to enhance ecological and aesthetic benefits.

In this work, the information from these sources was combined to develop quantitative indicators which estimate the level of impact of each measure on improving each co-benefit (Tables 3.2 a, b and c). Defining quantitative indicators is necessary in order to develop a ranking of GBI. This ranking will aid decision makers to choose among options.

Here, we introduce the indicators for the measures that are going to be used in the case study later presented. Some GBI are rather similar, but were differentiated in this case to consider different application possibilities. For instance, bio-retention areas and rain gardens are seen as a bigger and smaller scale of the same measure. While bio-retention areas can be applied in parks and other large green areas, rain gardens are more suitable for parking lots and transport corridors. Similarly, extensive and intensive green roofs could be grouped as only green roofs, but both give different co-benefits and their applicability depends on roof characteristics, among other factors (Woods-Ballard *et al.*, 2007).

3.4 STUDY AREA

Ayutthaya is located about 80 km north of Bangkok, the capital city of Thailand. It is an island of 7 km^2, surrounded by three rivers: Chao Phraya River, Pasak River, and Lop Buri River. The city includes a World Heritage Site (WHS) area, selected by UNESCO due to the importance of its historic and cultural sites. This WHS area covers around 289 ha (see Figure 3.6). Remaining areas are used for residential, educational, commercial, and government facilities.

The island has soils composed by clay and sand. Due to over exploitation of groundwater resources, aquifers and the overlying clay layer are under significant stress, leading to land subsidence in the region (Lorphensri et al., 2011). Bangkok exhibits the most critical situation, but the study area is only 80 km away and have also faced this issue. Other studies describe that WHS area has been also affected by this issue, putting under risk the cultural and architectural value of this area (Golub, 2014; Keerakamolchai, 2014).

The area is under high flood risk, caused by water levels of surrounding rivers and directly from heavy rainfall. Moreover, ground level is relatively low at approximately 4m Above Mean Sea Level (AMSL), contributing to the high risk of inundation (Keerakamolchai, 2014).

Table 3.2. (a) Impact of each measure on environmental benefits (data derived from Jia et al., 2013, Woods-Ballard et al, 2007, DEFRA, 2016, Berghage et al, 2009). (b) Impact of each measure on social benefits (data derived from Jia et al., 2013, Woods-Ballard et al, 2007, DEFRA, 2016). (c) Impact of each measure on economic benefits (data derived from Jia et al., 2013, DEFRA, 2016). The values of indicators are: 0 none/depreciable, 1 very low, 2 low, 3 medium, 4 high and 5 very high

(a)

ENVIRONMENTAL BENEFITS					
Name	Water quality	Groundwater recharge	Biodiversity and ecology	Temperature reduction	Air quality
Bio-retention area	4	2	4	3	2
Rain garden	4	1	3	2	2
Pervious pavement	5	3	1	3	0
Rain barrel	0	3	0	1	0
Detention pond	2	2	2	1	0
Retention pond	5	2	4	2	0
Green roof extensive	2	0	3	3	3
Green roof intensive	3	0	4	4	4
Bio-swale	4	2	3	3	2
Infiltration trench	5	4	1	2	2

(b)

SOCIAL BENEFITS			
Name	Amenity and aesthetics	Recreation and health	Food security
Bio-retention area	5	1	2
Rain garden	5	1	2
Pervious pavement	2	1	0
Rain barrel	0	0	0
Detention pond	3	3	1
Retention pond	4	3	3
Green roof extensive	3	2	0
Green roof intensive	4	4	5
Bio-swale	3	3	0
Infiltration trench	3	1	0

(c)

ECONOMIC BENEFITS			
Name	Rainwater harvesting	Pumping and treatment	Building energy	Real estate value
Bio-retention area	1	3	0	3
Rain garden	1	2	0	3
Pervious pavement	2	1	0	1
Rain barrel	5	4	0	2
Detention pond	3	5	0	2
Retention pond	5	4	0	4
Green roof extensive	0	0	3	2
Green roof intensive	0	0	4	3
Bio-swale	1	0	0	1
Infiltration trench	2	2	0	2

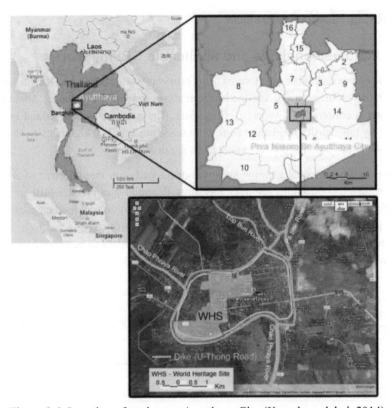

Figure 3.6. Location of study area Ayutthaya City (Keerakamolchai, 2014)

The current flood protection system of the island is formed by dykes at 5.3m AMSL to prevent fluvial flood. However, even under this protection, the city suffered from severe fluvial flood in 1995 and 2011. Additionally, due to the vicinity of the estuary and the Bay of Thailand, storm surges cause increased water levels in the area, due to impediment of discharges throughout the outfall (Golub, 2014). Furthermore, climate change is expected to contribute to the increment of flood risk in the area, due to higher number and intensity of tropical rains (Roachanakanan, 2013).

The main drainage network inside the island consists of canals and combined drainage system, which drain combine waste and storm water to pumping stations. GBI measures are proposed to improve stormwater management in the area, reducing the quantity of runoff conveyed by the drainage system. The measures considered were selected based on local characteristics and feasibility analysis.

3.5 RESULTS AND DISCUSSION

3.5.1 Land use analysis and applicable GBI measures

A land use analysis has been performed using previous works in the area (Keerakamolchai, 2014; Patiño Gómez, 2017), satellite image analysis and visits to the area. About half of the island is conserved as World Heritage Site. The other half of the island is mostly covered by low density residential zone, with some medium and high density residential zones, while education and government areas are dispersed.

An inventory of potential location sites was developed with the objective of analysing implementation possibilities for green measures. Potential sites to place GBI are defined according to land use in the island, covering a maximum of 30% of total case study area. These potential sites are well distributed over the total area (Figure 3.7). Through suitability analysis, it was defined which measures are feasible to be applied in each site type (Woods-Ballard *et al.*, 2007; UDFCD, 2010) (Table 3.3).

Table 3.3. Suitability analysis of GBI placement sites

Type of site	Green roof in.	Bio-retention	Rain garden	Green roof ex.	Bio-swale	Infiltration trench	Pervious pavement	Detention pond	Retention pond	Rain barrel
Non-flat roofs				▨						▨
Flat roofs	▨			▨						▨
Parking lots			▨		▨	▨	▨			
Transport corridors					▨	▨	▨			
Green spaces		▨			▨	▨			▨	
Parks and playfields		▨			▨	▨				
Water									▨	

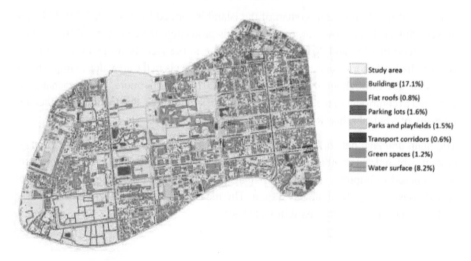

Study area
Buildings (17.1%)
Flat roofs (0.8%)
Parking lots (1.6%)
Parks and playfields (1.5%)
Transport corridors (0.6%)
Green spaces (1.2%)
Water surface (8.2%)

Figure 3.7. Location of different land uses in the study area

3.5.2 Identification of main co-benefits to be enhanced

To identify the most important co-benefits to be addressed in the area, two data sources were used. Firstly, the revision of previous studies developed in the area to identify local needs. Secondly, participatory analysis with stakeholders, using a questionnaire to collect data about stakeholders' perceptions. The aim of this analysis was to establish their opinion about what are the most important benefits that should be enhanced in the area.

Ayutthaya is an important touristic and cultural city, hence several previous studies have been performed in the area. Reviewing these works, the need of flood risk reduction was identified as a main concern (Vojinovic *et al.*, 2016a). Although the work presented here is about selection of strategies for flood risk reduction, the focus is on co-benefits enhancement. A parallel or posterior analysis must be done before making a final decision, to define the impact of different strategies on reducing flood risk.

Vojinovic et al. (2016b) studied different measures combining ecosystem services enhancement with flood modelling and cost-benefits analysis. Although this work focused mainly on fluvial floods reduction through centralised measures, it identified aesthetics and recreation as pertinent services for the study area. We also used the conclusions and recommendations of Golub (2014) and Keerakamolchai (2014) to identify the most important needs. Both applied participatory approaches to conclude that the focus should be on the enhancement of landscape by adding flood management measures, with a positive impact on benefits like biodiversity and ecology, amenity and aesthetics, tourism and water management. These studies also remark the problems in the WHS area due to land subsidence, which is originated on groundwater table depletion. Consequently, groundwater recharge is seen another important benefit that should be enhanced.

In 1997, the Study of Public Works Department, Ministry of Interior, defined flood protection measures for Ayutthaya. These measures were based on engineering solutions and cost benefit analysis, without considering public perception, environment or aesthetics. Consequently, the proposed solution was controversial, creating arguments with the community, which at the end resisted the plan (Keerakamolchai, 2014). This experience serves as an example of the importance of including stakeholders' perception when planning the implementation of measures. Besides, participatory planning processes are seen as useful tools when stakeholders perceive opposing main goals, helping them to accept plans that aim a variety of goals, such as the multiple benefits delivered by GBI (Cheng *et al.*, 2017).

This area has been the focus of several workshops and consultation meetings on this topic (Golub, 2014; Keerakamolchai, 2014; Vojinovic *et al.*, 2016a, 2016b). In order to avoid over questioning or exhaustion of stakeholders, we chose to apply a simple and short questionnaire (see Appendix B). In addition, this method avoids attendance problems and

47

allows to reach a broader diversity of people, such as inhabitants and tourists. The questionnaire included fifteen questions, which were developed based on CIRIA works (Woods-Ballard *et al.*, 2007; CIRIA, 2013). The benefits included were divided in the three categories described in section 3.3.1: environmental, social and economic benefits.

The questionnaire was answered by 42 stakeholders from different backgrounds: public authorities, international agencies and groups, private sector, citizens and visitors. A diversity of methods to apply the questionnaire were used. The main method used was a Google form that was filled and submitted online. Another method used was one-to-one interviews, applied in the case of commercial owners, citizens and tourists around the study area (for further details see Patiño Gómez, 2017). The main groups that answered the survey were people living in the area (inhabitants), researchers, tourists and people from local government (see Figure 3.8).

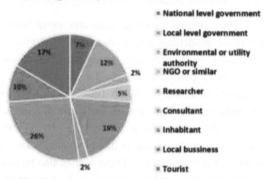

Figure 3.8. Type of stakeholders that answered the questionnaire

The first questions were related to the most important benefit that should be enhanced in each category (Figure 3.9). By analysing the results, it was concluded that these benefits are: biodiversity and ecology (with 38% of answers), amenity and aesthetics (with 50% of answers) and rainwater harvesting (with 48% of answers). Previous studies obtained similar results regarding the benefits that should be improved in the area (Golub, 2014; Keerakamolchai, 2014).

The remaining questions focused on the importance of each benefit separately (Figure 3.10). In this case the most important benefits for each category were maintained with respect to the ones shown in Figure 3.9. However, the second most important benefits were different in the cases of environmental and economic benefits. This is important because the objective here was to choose the first and second preferred benefits as key co-benefits for the case under study.

Water quality improvement was the benefit chosen as second most important when participants were asked to compare among environmental benefits, with 29% of answers (Figure 3.9-a). However, when participants were asked to evaluate each benefit separately, water quality improvement falls to the fourth place. This is because 59% of the participants classified as "high" the relevance of air quality improvement and only 31% chose water quality improvement (Figure 3.10 a to e).

Regarding economic benefits, the second most important benefit when compared among them was reduction of building energy consumption, with 24% of answers (Figure 3.9-c). However, the second most important benefit when they had to evaluate one by one, was reduction of stormwater pumping and treatment, with 76% of answers choosing it as high importance benefit, while 67% of respondents chose reduction of building energy consumption (Figure 3.10 h to k).

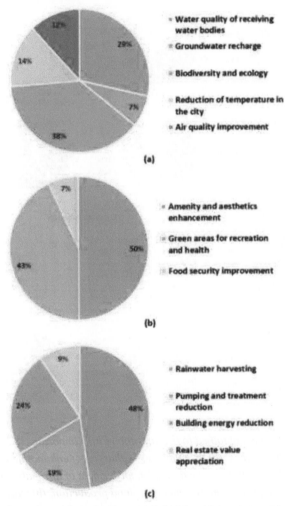

Figure 3.9. Most important benefits from stakeholders' interviews: (a) environmental benefits, (b) social benefits, (c) economic benefits

In this work, we consider that the direct comparison among benefits gives a clearer understanding and a more reliable result. Consequently, the environmental benefit selected as second most important was the improvement of water quality of receiving bodies. In the case of economic benefits, reduction of energy consumption in buildings was chosen as the second most important benefit.

In summary, biodiversity and ecology and water quality of receiving bodies were identified as key environmental benefits. Amenity and aesthetics enhancement, and green areas increase were selected as the main social benefits. In the case of economic benefits, rainwater harvesting and building energy reduction were identified as the most important.

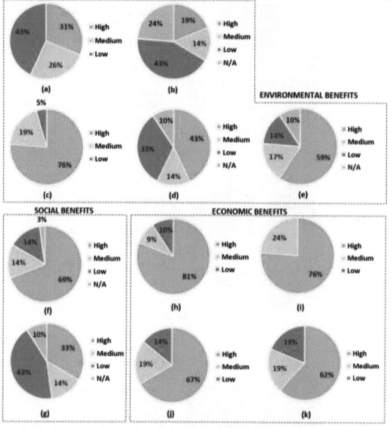

Figure 3.10. Answers obtained to the question of how important each benefit is: (a) pollution of water bodies, (b) groundwater recharge, (c) enhancement of biodiversity and ecology, (d) heat stress reduction, (e) improvement of air quality, (f) increment of green spaces and amenity improvement, (g) food production, (h) water harvesting, (i) reduction of water pumping and treatment, (j) reduction of buildings energy consumption, (k) real estate vale increment

Another analysis performed focused on the answers given by different stakeholder's types. The objective was to understand how different actors perceive different necessities. Stakeholders were divided in three groups. The first one represents general public, including inhabitants, local commercial owners and tourists. The second one represents policy makers and is formed by people from national government and local government levels. The last one includes people from the scientific community, such as researchers, consultants and non-governmental organizations (NGO).

Among the three actors included into the general public group, inhabitants and local business owners selected the same benefits already identified as first and second options from the previous analysis (Figure 3.11-a). While tourists had differences when choosing the second most important benefits, selecting food security and real estate value as second option for social and economic benefits respectively.

Regarding the policy makers, both subgroups chose the previously identified first and second most important benefit in the cases of environmental and social benefits (Figure 3.11-b). However, in the case of economic benefits, people from national government did not show a preference, while at the local government level the second preferred option was pumping and treatment reduction.

Finally, results from the third group showed that consultants and NGO people chose biodiversity and ecology improvement as the only important environmental benefit, while researchers selected temperature reduction as the most important environmental benefit (Figure 3.11-c). In the case of social benefits, both groups selected the already identified first and second most important benefits. Regarding economic benefits, researchers selected the key benefits defined from the general analysis, while consultants and NGO people indicated pumping and treatment reduction as the most important benefit.

These results confirm how the common practice of making decision from a unilateral point of view, by the scientific community or policy makers, does not allow to consider preferences of residents or visitors with a much more local perspective. For instance, in this case, if general public perceptions are not considered temperature reduction could be chosen as key environmental benefit in place of water quality improvement. Or pumping and treatment reduction could be perceived as more important than building energy savings in the case of economic benefits. This analysis shows the importance of including different types of stakeholders when making decisions, and in particular the inclusion of residents and visitors to consider local perceptions.

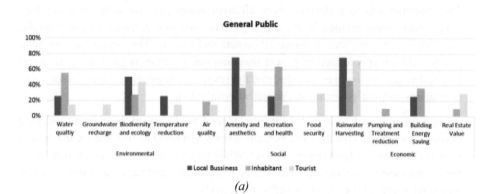

(a)

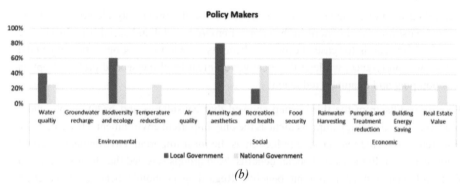

(b)

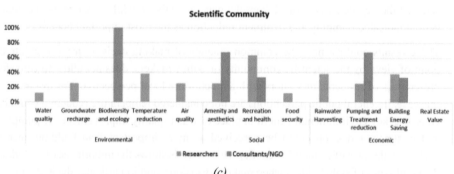

(c)

Figure 3.11. Answers to the questionnaire from different stakeholders' types: (a) General public: local business owners, inhabitants and tourists, (b) Policy makers: people form local and national government, (c) Scientific community: researches and consultants and NGO people.

3.5.3 Definition of the most effective combinations of GBI

Based on the most important co-benefits identified from the analysis of stakeholders' perceptions, and the quantitative indicators defined in Section 3.3.2, a ranking of GBI measures was developed (Table 3.4). The ranking considers only the key co-benefits identified by the analysis of stakeholders' answers. For each measure, the indicators for each one of these co-benefits were added to obtain a final score. According to this ranking, the most convenient drainage GBI measures to enhance the selected co-benefits in Ayutthaya are: retention pond, intensive green roof, bio-retention area, rain garden, and bio-swale.

Table 3.4. Ranking of measures according to impact on selected co-benefits

GI measures	Selected co-benefits						Total benefits score
	Biodiversity and ecology	Water quality enhancement	Amenity and aesthetics	Recreation and health	Rainwater harvesting	Building energy reduction	
Retention pond	4	5	4	3	5	0	21
Green roof intensive	4	3	4	4	0	4	19
Bio-retention area	4	4	5	1	1	0	15
Rain garden	3	4	5	1	1	0	14
Bio-swale	3	4	3	3	1	0	14
Green roof extensive	3	2	3	2	0	3	13
Detention pond	2	2	3	3	3	0	13
Infiltration trench	1	5	3	1	2	0	12
Pervious pavement	1	5	2	1	2	0	11
Rain barrel	0	0	0	0	5	0	5

Combining the results obtained from this ranking with results from the land use and measures suitability analyses, the preferred measures for each type of land use were defined (Table 3.6-a). For each land use type, the most preferred and second preferred GBI measures are considered.

Through these results, we identified potential measures to apply in each land use type of Ayutthaya. The most convenient GBI measures to apply at building level are green roofs, preferably intensive green roofs, applicable in flat roof buildings. In parking lots and transport corridors, the most effective GBI to achieve the selected co-benefits are bio-swale and rain garden. In the cases of parks and playfields and green places, bio-retention area appears as the most adequate GBI to be used. Finally, retention pond is recommended as effective GBI measures to be applied in water surfaces.

In this work we focused on the identification of key co-benefits as the central aspect to select flood risk reduction strategies. Nevertheless, the capacity of GBI measures to reduce flood risk should be included in the analysis from the very beginning of decision making processes. To achieve this, a qualitative analysis of this aspect was introduced. The selection of a qualitative analysis is based on its simplicity, since in this case the objective was to see how preferred measures change when this aspect was introduced.

However, a flood risk reduction analysis should be based on a more complex study including hydrodynamic modelling.

Woods-Ballard *et al.* (2007) describes in qualitative terms the capacity of different GI measures to reduce peak flow and runoff volume (Table 3.5). According to this information, the most effective measures to reduce flood risk, among the ones considered for this case, are pervious pavement, rain barrels and detention and retention ponds. However, not all these measures appear as preferred to enhance the key co-benefits identified for this case study.

Table 3.5. Flood risk reduction capacity of selected GBI measures (Woods-Ballard et al., 2007)

GI measures	Flood risk reduction capacity	
	Peak flow reduction	Volume reduction
Green roof intensive	Medium	Medium
Bio-retention area	Medium	Medium
Rain garden	Medium	Medium
Green roof extensive	Medium	Medium
Bio-swale	Medium	Medium
Pervious pavement	Good	Good
Infiltration trench	Medium	Medium
Detention pond	Good	Low
Retention pond	Good	Low
Rain barrel	Good	Good

If flood risk reduction capacity is included into the analysis as a main objective to select GBI measures, pervious pavements should be included as preferred measures to apply in the cases of parking lots, parks and playfields and transport corridors. Likewise, detention ponds should be considered in the cases of green spaces and parks and playfields, and rain barrels for the cases of flat and non-flat roofs. Table 3.6-b presents the addition of these options to the most preferred measures previously identified for co-benefits enhancement.

The results presented in Table 3.6 show that measures which were not preferred when only co-benefits were considered, appear as favourite for flood risk reduction, and vice versa. Consequently, if only flood reduction capacity is considered, as is common during decision making for stormwater management, the improvement of co-benefits would be neglected. Also, if only co-benefits are considered, the reduction of flood risk would be minimal. This demonstrates the importance of considering both objectives. In the case presented here, decision makers should consider the mixture of measures marked in green and blue in Table 3.6-b to achieve sustainable strategies to decrease flood risk and enhance co-benefits.

When these results are analysed from an ecosystem services point of view, it can be observed that provisioning, regulating and cultural services are improved through the implementation of the measures selected. Provisioning of native plants, birds and insects can be improved implementing green roofs, bio-retentions, rain gardens and bio-swales. Regulating and maintenance services such as baseline flow maintenance, flood protection, combined sewer overflows and air quality improvement, are achieved using infiltration and vegetated measures. Water pollution removal is reached using bio-retentions areas, while mediation of smell, noise and visual impacts are improved using green measures in transport corridors. Pollination and seeds dispersal, creation of habitats for plants and animals, climate regulation through carbon sequestration and local temperature reduction, are achieved implementing green infrastructures such as green roofs, rain gardens, bio-retention areas and bio-swales. Finally, cultural services such as physical interactions watching birds and plants, as well as intellectual interactions with scientific and educational opportunities, and aesthetic improvements of heritage areas are enhanced applying the different measures selected for this case.

Table 3.6. (a) Preferred measures according to co-benefits for each land use category (preferred: dark green; second preferred: medium green); (b) Preferred measures according to co-benefits (preferred: dark green; second preferred: medium green) and flood risk reduction (preferred: blue) for each land use category.

(a)

Type of site	Green roof in.	Bio-retentio	Rain garden	Green roof ex.	Bio-swale	Infiltration trench	Pervious pavement	Detention pond	Retention pond	Rain barrel
Non-flat roofs										
Flat roofs										
Parking lots										
Transport corridors										
Green spaces										
Parks and playfields										
Water										

(b)

Type of site	Green roof in.	Bio-retention	Rain garden	Green roof ex.	Bio-swale	Infiltration trench	Pervious pavement	Detention pond	Retention pond	Rain barrel
Non-flat roofs										
Flat roofs										
Parking lots										
Transport corridors										
Green spaces										
Parks and playfields										
Water										

3.6 CONCLUSIONS

This chapter presented a methodological framework for selection of green-blue infrastructures combinations based on the improvement of co-benefits, besides the reduction of flood risk. The achievement of co-benefits was introduced here as a consequence of green infrastructure implementation, which also allows ecosystem services enhancement and the consequent human well-being.

The framework presented combines land use analysis with the identification of locally needed co-benefits, achieved through stakeholders' survey. As new concept, the framework includes the consideration of key co-benefits into decision making processes to select flood risk reduction strategies. Furthermore, the method takes into account participatory planning and stakeholder's perceptions analysis to identify these key benefits.

The framework was applied to a case study in Ayutthaya, Thailand. The results obtained showed the importance of participatory planning processes, which help decision makers to develop sustainable solutions based on local necessities, particularly in cases where stakeholders have different objectives. In this case, different stakeholder's types showed different perceptions about key benefits to be enhanced, confirming that unilateral decision making processes from policy makers or scientific community, could focus on different benefits than the ones identified as most important by local residents.

Moreover, the results showed the importance of considering co-benefits when designing strategies to reduce flood risk. Significant differences are observed when comparing preferred measures considering only key co-benefits enhancement and considering only flood risk reduction. This remarks the importance of taking into account both objectives from the very beginning in decision making processes to achieve sustainable flood reduction strategies. Using this approach, a complete range of benefits achievable combining different green infrastructures can be capitalized.

The methodology presented here does not pretend to be conclusive. Several steps in this method were based on qualitative analysis, in particular the evaluation of measures flood risk reduction. Future work is needed to combine the framework presented in this work with a complete flood risk reduction analysis of green measures combinations.

4

ASSESSING THE CO-BENEFITS OF GREEN-BLUE-GREY INFRASTRUCTURE

Green-blue infrastructures in urban spaces offer several co-benefits besides flood risk reduction, such as water savings, energy savings due to less cooling usage, air quality improvement and carbon sequestration. Traditionally, these co-benefits were not included in decision making processes for flood risk management. This chapter presents a method to include the monetary analysis of these co-benefits into a cost- benefits analysis of flood risk mitigation measures. This approach was applied to a case study, comparing costs and benefits with and without co-benefits. Different intervention strategies were considered, using green, blue and grey measures and combinations of them. The results obtained illustrate the importance of assessing co-benefits when identifying best adaptation strategies to improve urban flood risk management. Otherwise green infrastructure is likely to appear less efficient than more conventional grey infrastructure. Moreover, a mix of green, blue and grey infrastructures is expected to result in the best adaptation strategy as these three alternatives tend to complement each other. Grey infrastructure has good performance at reducing the risk of flooding, whilst green infrastructure brings in multiple additional benefits that grey infrastructure cannot offer.

Based on Alves A., Gersonius B., Kapelan Z., Vojinovic Z., Sanchez A. (2019). "Assessing the Co-Benefits of green-blue-grey infrastructure for sustainable urban flood risk management". Journal of Environmental Management, 239, 244-254, doi: 10.1016/j.jenvman.2019.03.036.

4.1 INTRODUCTION

Currently, most people in the world live in cities and urban population is expected to continue growing in the future (United Nations, 2014). We need to enhance liveability and sustainability in cities, ensuring that urban spaces are safe and attractive places for living and working. However, climate change and urbanisation are putting this at risk, through problems such as increasing flood risk, heat stress, wáter shortages and air pollution (IPCC, 2012). Green-blue infrastructure (GBI) offer a multifunctional approach which can reduce vulnerability and increase resilience in front of these multiple threats (European Commission, 2012a).

Traditionally, flood management was focused on grey or traditional solutions, such as pipes. Nowadays, it is understood that this approach offers low sustainability while GBI provide numerous complementary benefits (Vojinovic, 2015). However, in practice, the use of GBI as an option for climate adaptation is still shadowed by the use of grey infrastructure. This is because these technologies are often evaluated from a single goal perspective, such as storm water management (Engström *et al.*, 2018). But it is through a comprehensive analysis of their multiple benefits that the complete net-benefits of GBI can be understood (Foster *et al.*, 2011). The secondary benefits, besides flood management, are called here co-benefits and are the positive side effects for people and the environment obtained from GBI application.

Economic valuation, including all relevant costs and benefits, is an important tool to support decision-making when planning GBI, particularly when comparing different types of infrastructures investment options (Wild *et al.*, 2017). A frequently used method to estimate the efficiency of projects is cost-benefit analysis (CBA), being the project attractive if the benefits are higher than the costs. In the case of flood management, the comparison is typically between the cost of measures to increase safety and the reduction of expected damages. This method offers significant rational information for decision makers when choosing among different solutions (Jonkman *et al.*, 2004).

Despite the challenges to monetise co-benefits it is recognised that monetary valuation helps to raise policy makers' awareness regarding the economic importance of these associated benefits (Saarikoski *et al.*, 2016; Chenoweth *et al.*, 2018). Very few works use net present value assessment over the lifespan of measures to compare sustainable and traditional flood management strategies from a holistic perspective, this is including the multiple benefits offered by GBI. For instance, (Urrestarazu Vincent *et al.*, 2017) shown that the economic feasibility of GBI is significantly improved if multiple benefits are considered. A similar result was obtained by and (Ossa-Moreno *et al.*, 2017) but working only with sustainable measures.

There is still the need to better understand how costs and benefits change when GBI are combined with grey solutions (Foster *et al.*, 2011). This chapter addresses this gap comparing the economic viability of green, blue and grey flood risk reduction strategies, focusing on the combination of different measures. A method to perform monetization of co-benefits and to include this into a cost-benefits analysis is presented. Besides the primary benefit of flood risk reduction, several secondary benefits, or co-benefits, are considered in this work. An analysis with and without co-benefits consideration is applied in a case study comparing different strategies combining green, blue and grey measures. The work performed falls into the framework of EC-founded PEARL (www.pearl-fp7.eu; Vojinovic, 2015) and RECONECT (www.reconect.eu) projects.

4.2 METHODOLOGY

In a traditional flood management measures assessment, only the primary benefit of flood damage reduction is considered. In this case, the assessment includes several secondary benefits (i.e. co-benefits), in addition to flood damage reduction. For instance, heat stress reduction, air quality improvement and water savings. These co-benefits are associated with the application of green and blue measures. Figure 4.1 summarises the whole methodological process.

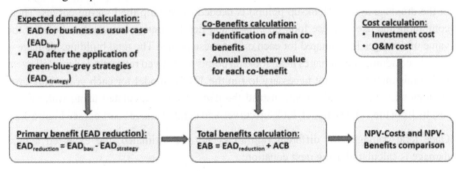

Figure 4.1. Methodology for total benefits and costs comparison

The primary benefit, seen here as the benefit obtained from flood damage reduction, is estimated by expected annual damage (EAD). EAD is the probabilistic expected flood damage cost per year for all possible flooding events and is expressed in monetary terms (Delelegn *et al.*, 2011). The flood damage reduction is then calculated as the difference between EAD in the case of business as usual (without measures), and EAD after the application of flood reduction measures.

CBA requires the quantification of all costs and benefits in monetary terms. This is achieved here calculating the monetary value per year of every relevant co-benefit obtained from GBI application (besides the primary benefit). The addition of EAD reduction and annual co-benefits (ACB), both due to the application of a specific measure or measures combination, is the value of total expected annual benefits (EAB).

Meanwhile, the calculation of costs considers investment and maintenance costs of every applied measure. Once total benefits and costs are estimated, both are converted to the net present value (NPV). This allows the comparison of these figures, seen as present values of costs and benefits, and to stablish which is higher over the project lifespan.

4.2.1 EAD calculation

An extensively used method to calculate flood damages comprises the use of 1D-2D hydrodynamic models and depth-damages curves. In practice, the damages of flooding are influenced by several factors, but usually water depth is the most influential factor in the case of small scale urban catchments (Delelegn *et al.*, 2011).

Due to the requirements on computational resources and time of 1D-2D models, in this case we estimate damages using a surrogate model. A surrogate model is a model that approximates a more complex and too computationally expensive model, allowing faster approximations (Udoh and Wang, 2009). In this case, the surrogate model emulates the original 2D one and is composed by a much simpler 1D-1D model and look-up curves.

The relation between water depth and the number of affected buildings in different points of the drainage system (look-up points) is needed to develop the look-up curves. Then, combining the use of these look-up curves and depth-damages curves for the area, the value of damages is estimated for each one of these points. The total buildings damage is calculated adding the damages in all look-up points for a given return period rainfall event. Using this method it is not necessary to run the 1D-2D model for each green-blue-grey strategy and each rainfall event, instead the damages are calculated using water depth results from the 1D-1D model, look-up curves and depth damages curves (see Figure 4.2).

Once total damages for different rainfall events are estimated, the expected annual damage is calculated using next equation:

$$EAD = \sum_{i=1}^{n} \left(\frac{D_i + D_{i+1}}{2} \right) * \left(\frac{1}{RP_i} - \frac{1}{RP_{i+1}} \right) \tag{4.1}$$

where D_i is the damage corresponding to the event of return period RP_i, and n is the number of return periods considered. The damage D is calculated as the addition of damages in the different look-up points.

To apply the concept of EAD, the return periods considered should cover a range going from frequent and not so damaging events, to a very rare event. Moreover, n should be as large as possible in order to have a good scatter of events (Delelegn *et al.*, 2011).

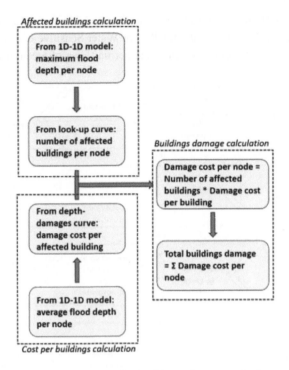

Figure 4.2. Process for buildings damage calculation

4.2.2 Co-benefits calculation

The first step to estimate co-benefits is the identification of locally relevant benefits and the applicable measures to achieve these benefits. In this work, this is accomplished through a multi criteria method for measures selection (Alves *et al.*, 2018a). This method is based on questions regarding local characteristics and preferred benefits. A ranking of green, blue and grey measures is built using the answers given by decision makers to these questions. From this ranking, different combinations of measures are selected for further analysis.

After the selection of measures, and the identification of benefits from them provided, the next step is the economic value calculation of these benefits. This is achieved understanding the relation between impacts on the environment and the consequent human welfare. A good description of these interactions is provided by (Defra, 2007) through the concept of impact pathways (see Figure 4.3).

Always when possible, the economic values of benefits are estimated based on local data, for instance energy and water prices. When local data is not available, general information form literature review is used. There are several published works on co-benefits values estimation, such as Horton *et al.* (2016), Center for Neighborhood Technology (2010), NYC Environmental Protection (2013), CRC for Water Sensitive Cities (2016), among other. The values of co-benefits are calculated per unit area of green infrastructure and per year.

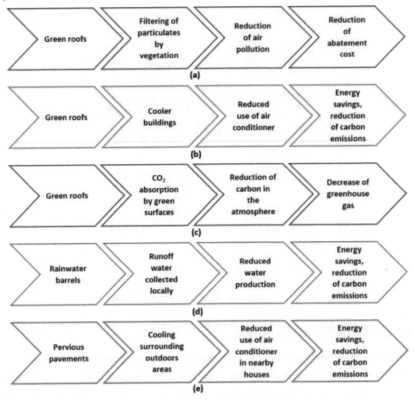

Figure 4.3. Conceptual analysis of five co-benefits monetization: (a) air quality, (b) buildings temperature reduction, (c) carbon sequestration, (d) rainwater harvesting, (e) heat stress reduction (adapted from Horton et al. (2016))

4.2.3 Costs calculation

Costs calculation is based on local prices and literature review. Afterwards, the obtained values are compared with values from other works for validation. In this work we consider investment and maintenance costs through the lifespan period of each infrastructure. All amounts are converted to the same year valuations using the consumer price index (CPI). Moreover, all values are converted to present values using NPV for the lifespan period of the measures. Next equations show these two conversions.

$$Value_{year\ A} = Value_{year\ B} * \left(CPI_{year\ A}/CPI_{year\ B}\right) \qquad (4.2)$$

$$NPV = \sum_{i=1}^{m} \frac{Value\ per\ year}{(1 + dr/100)^i} \qquad (4.3)$$

4.3 RESULTS

4.3.1 Study area description

The methodology here presented was applied in a case study in the Dutch side of Sint Maarten Island, located in the Caribbean region. This part of the island covers an area of approximately 3380ha. Elevation ranges from near sea level at the southern end, to hilly areas with until 380 mat the northern borderline. Stormwater catchments and streams have several characteristics contributing to severe flood related impacts. For instance, urban areas are situated on low-lying zones, with not good stormwater drainage infrastructure. Besides, streets in residential areas are usually narrow allowing very limited further enlargement of stormwater channels (see also Vojinovic and van Teeffelen, 2007).

The catchment selected for this study is called Cul De Sac, which is one of the most vulnerable areas to flooding. This catchment has an area of 509ha and the land use is predominantly residential, with some dispersed commercial areas in the lower part. During small rainfall events usual impacts include inconveniences such as disruption to transportation systems. However, heavy rainfall causes large-scale flooding with damages to residential and commercial buildings (UNDP, 2012).

The local currency is Netherlands Antillean Guilder. However, in practice USD, Euros and the local currency are accepted. Regarding the data used in this work, damage curves are given in Euros, while other information for costs and co-benefits calculations is either in Euros or USD. In this work all values are given in Euros.

4.3.2 Measures selection and benefits screening

The multi-criteria analysis introduced by Alves *et al.* (2018) was used to choose the measures to be studied in this work. To apply this method a questionnaire was answered by local decision-makers. The questionnaire was focused on the obtention of local physical characteristics data, such as soil type and water table depth. The respondents also had to define local preferences regarding co-benefits, for instance choosing which benefit is more important between water quality and liveability. The answers were used to develop a ranking of green-blue-grey measures. From this ranking, five measures were selected. Green roofs, pervious pavements and rainwater barrels (as green infrastructure) were chosen to be applied in the flat and more urbanized area of the catchment. Open detention basins (as blue infrastructure) and pipes (as grey infrastructure) were selected to manage runoff from steep areas and to increase the capacity of the existent drainage system, respectively. This analysis considers the existing channels system working properly for a two years return period storm.

Next, we analysed the main co-benefits obtained through the selected measures and their importance for the case under study. Green roofs offer several benefits besides runoff reduction, such as thermal insulation of buildings, air pollution reduction and carbon sequestration, as well as longer lifespan than traditional roofs (Kosareo and Ries, 2007; Rowe Bradley, 2011; Bianchini and Hewage, 2012). Buildings insulation is crucial in this island, which has high temperatures and consequently high energy consumption for cooling. Furthermore, energy savings is very important since the island has expensive energy production. The energy production depends on imported fossil fuel which implies high carbon footprint and air pollutants emission. Additionally, the island has one of the highest regional electricity prices and energy consumption rates (Radjouki and Hooft Graafland, 2014).

The installation of rainwater harvesting barrels at household level is a useful measure which allows the reduction of drinking water consumption. This is an important benefit in an area where water production and cost have been increasing notoriously in the last 10 years (Centrale Bank Curaçao en Sint Maarten, 2017). The production of water in the island is based on reverse osmosis, an expensive technology which implies high energy consumption (Elimelech and Phillip, 2011). Moreover, the Dutch part of the island experiences water shortages during peak consumption hours (European Commission, 2012b).

Regarding the installation of pervious pavements, this measure allows urban cooling through lower reflection and higher evaporation (Foster *et al.*, 2011). This reduction of surface temperature can reach between 8 and 3 Celsius degrees during day and night respectively (Charlesworth, 2010). The benefits obtained from this reduction are cooling energy savings, as well as associated carbon dioxide and air pollutants reduction (USEPA, 2012). Temperature reduction is especially important for an area with tropical weather and high average temperatures, where the increment of energy consumption can reach 2–4% per each Celsius degree of higher temperature (Akbari *et al.*, 2001; Santamouris, 2014).

In a previous work, open detention ponds (ODP) were identified as applicable flood management alternatives in Cul De Sac catchment (UNDP, 2012). This was reaffirmed through the measures selection process applied for this case. As a result, we analysed the application of several of these structures in available spaces upstream the flat and more densely populated area. Although multi-functionality can be considered for this measure, allowing co-benefits as recreation and liveability enhancement, this benefits were not included for the present study.

Finally, pipes were chosen as grey measure to increase the capacity of the existent drainage system in the catchment. This system is composed by open channels with limited capacity to convey the excess of rainfall runoff. Several of these channels are located in narrow streets which do not allow their enlargement (UNDP, 2012). A single pipe is planned to follow the main channel path from the mid area of the catchment until its discharge.

4.3.3 Damage calculation

A 1D-1D flood estimation model was developed in Storm Water Management Model (SWMM), version 5.0 (Rossman, 2010). It includes the main drainage channels in the area, linked to streets and surfaces which represent the floodplain. This model was calibrated with results from an existent 2D model developed in DHI software Mike FLOOD (UNDP, 2012). Values of water depths from the previously mentioned 1D-2D model were registered in several points of the drainage system (look-up points) to develop look-up curves (Water Depth-Buildings Damaged curves). The construction of the curves was done obtaining the number of affected buildings for different flood depths in those points.

One look-up curve was built for each look-up point (represented as a node in the 1D-1D model). Figure 4.4 shows an example of how to construct look-up curves. It presents a node location, flooding surfaces for different return periods, buildings in the area, and the resultant curve for that node. Once the look-up curves were constructed, the procedure used to calculate damages was as follows. Firstly, changes were made to the 1D-1D model, adding the selected measures for flood risk management. Secondly, the model was run and maximum water depths in each node were registered. With this data, the number of affected buildings per node were obtained from the curves previously built.

To estimate damage costs in buildings, Water Depth – Damage curves for residential and commercial buildings in Sint Maarten were used, these curves were obtained from Huizinga *et al.* (2017). Combining the results of number of buildings affected and depth-damage curves for the area, residential and commercial damages were estimated for different return period rainfall events. Other damages considered in this study are infrastructure damage and transport damage. Again, values from Huizinga *et al.* (2017) and average water depths per node from the 1D-1D model were used. Table 4.1 shows damage values in the case of business as usual, this is without the application of flood management measures.

The calculation of EAD and its NPV is shown in Table 4.2. Notice that this EAD corresponds to the current situation, and is the maximum value obtained. After applying measures, it is expected the reduction of flooding and consequently the reduction of damages.

Table 4.1. Calculated damages for different return period rainfalls

Damages (1x10^6 €)	Return period				
	5	10	20	50	100
Residential	2.8	5.9	13.2	19.7	23.7
Commercial	0.7	0.9	1.3	1.6	1.8
Infrastructure	0.8	1.5	2.6	3.5	3.9
Transport	0.1	0.5	1.7	3.3	4.1
Total	**4.4**	**8.8**	**18.8**	**28.1**	**33.5**

Table 4.2. EAD calculation for current situation and NPV of EAD in a period of 30 years

Return period	Event frequency	Damage per event (1x10^6 €)	EAD (1x10^6 €)
2	0.5	0	
5	0.2	4.4	0.7
10	0.1	8.8	0.7
20	0.05	18.8	0.7
50	0.02	28.1	0.7
100	0.01	33.5	0.3
		Total EAD	**3.0**
		EAD NPV$_{30}$	**46.6**

The period used for NPV calculation was 30 years with 5% discount rate (International Monetary Fund, 2016). This period is considered appropriated without replacement of measures when working with green infrastructure. For instance, different authors establish between 30 and 55 years of lifespan for green roofs (Porsche and Kohler, 2003; Kosareo and Ries, 2007; Rowe Bradley, 2011; Bianchini and Hewage, 2012; Claus and Rousseau, 2012). In the case of pervious pavements, life time before clogging is estimated between 15 and 25 years (Pezzaniti *et al.*, 2009; USEPA, 2012; Al-rubaei *et al.*, 2013; Yong *et al.*, 2013). Regarding the discount rate, several studies on this topic were considered to validate this discount rate value. Discount rates in these works vary between 2 and 8% when working with green measures and flood damage mitigation (Jonkman *et al.*, 2004; Carter and Keeler, 2008a; Bianchini and Hewage, 2012; Claus and Rousseau, 2012).

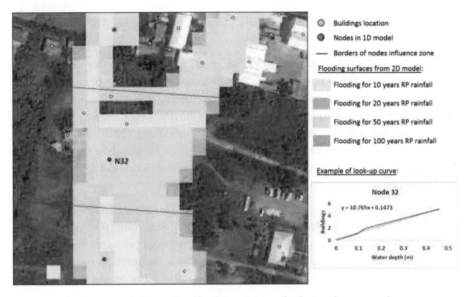

Figure 4.4. Example of look-up curve calculation for one node

4.3.4 Co-benefits calculation

Green Roofs

The main direct benefit obtained from green roofs was energy savings due to building temperature reduction. To calculate this benefit, we applied the method presented by (Center for Neighborhood Technology, 2010). This method provides a simple estimation of building energy savings, seeing green roofs as an insulation and assuming that a reduction in heat flux produces direct energy savings (see Equation 4.4). The result for this benefit is 1.61 €/m²/year.

$$Energy\ savings\ (kWh/m^2/y) = C \times \left[\left(\frac{1}{R_{conv\,roof}}\right) - \left(\frac{1}{R_{green\,roof}}\right)\right] \times \frac{24hrs}{day} \times 0.00315 \quad (4.4)$$

where C=annual number of cooling degree days (°F*days); $R_{conv\,roof}$ = thermal resistance for conventional roofs (11.34 SF*°F*hrs/BTU); $R_{green\,roof}$ = thermal resistance for green roofs (23.4 SF*°F*hrs/BTU); 0.00315 is factor to convert from BTU/SF to kWh/m²; BTU = British thermal units.

Annual cooling degree days is an estimation of how hot the climate is and is used to calculate the energy needed to keep buildings cool. This value is calculated as the difference between a balance temperature and the mean daily temperature, and adding only positive values over an entire year. In this case the estimation of annual cooling savings was done considering four months of 27 celsius degrees as an average

(Meteorological Department St. Maarten, 2018), with 20 celsius degrees as balance temperature.

Two indirect benefits are obtained from energy savings because in this case energy is obtained from fossil fuel power plants. These benefits are reduction of carbon dioxide (CO_2) emissions and improvement of air quality. Regarding the later, two pollutants were considered, nitrogen dioxide (NO_2) and sulphur dioxide (SO_2). Again, we used the methods presented by Center for Neighborhood Technology (2010) to calculate the quantities of pollutants avoided and their economic values. Results show that savings due to air quality improvement are 0.08 €/m^2/year; while savings due to CO_2 reduction are 0.17 €/m^2/year.

The installation of green roofs also has a direct impact on air quality improvement. In this case, reductions of four pollutants were considered: NO_2, SO_2, ozone (O_3) and particulate matter (PM). The quantities of air pollutants directly removed per square meter of green roof and per year, as well as the economic value of these pollutants, are average values provided by Center for Neighborhood Technology (2010). The benefit obtained is 0.5 €/m^2/year.

Another direct impact of green roofs is CO_2 sequestration. Although there are other types of greenhouse gases contributing to climate change, CO_2 is the one most directly affected by green infrastructure (Center for Neighborhood Technology, 2010). The annual amount of carbon sequestered is calculated as the total area of green roofs times the average annual amount of carbon sequestered per unit area of green roof. The range of carbon sequestered per area of green roof considered is 162–168 g C/m^2 (Getter et al., 2009). The direct benefit due to carbon sequestration is 0.03 €/m^2/year.

Finally, the implementation of green roofs increases roofs longevity because the membrane is protected from weather conditions by the soil layer (Kosareo and Ries, 2007). A conventional roof has a lifespan of between 10 and 20 years (Kosareo and Ries, 2007; Claus and Rousseau, 2012). While the lifespan of green roofs is expected to be between 40 and 55 years (Carter and Keeler, 2008b; Rowe Bradley, 2011; Bianchini and Hewage, 2012; Claus and Rousseau, 2012). Values of re-roofing are established between 92 and 160 USD/m^2 (Montalto et al., 2007; Bianchini and Hewage, 2012) . In this work, an average value of 108 €/m^2 applied every 20 years was considered as the investment avoided for roof retrofitting.

More details about these calculations are presented in Table 4.3. It shows savings and prices for all the benefits obtained from green roofs installation. For instance, values of energy saved and price (National Renewable Energy Laboratory, 2015) due to building insulation are presented. Concerning the calculation of carbon dioxide reduction due to energy savings, emissions due to oil based electricity production were estimated as 1.616 lb CO_2/kWh (WNA, 2011). An average carbon price of 0.02 €/lb CO_2 was assumed

(Center for Neighborhood Technology, 2010). The accumulation of direct and indirect co-benefits related to green roofs gives a value of 2.91 €/m^2/y, and a NPV over 30 years of 447823 € per hectare of green roof installed.

Table 4.3. Calculation of green roofs annual benefits

Building temperature: energy savings (direct)

Cooling savings (KWh/m^2)	5.38	Energy price (€/KWh)	0.3

Air quality due to energy savings (indirect)

Electricity production by fossil fuel produces the emission of NO$_2$ and SO$_2$

NO$_2$ avoided (g/KWh)	0.88	NO$_2$ value (€/g)	0.0063
SO$_2$ avoided (g/KWh)	2.39	SO$_2$ value (€/g)	0.0039

Carbon reduced due to energy savings (indirect)

Electricity production by fossil fuel produces the emission of CO$_2$

CO$_2$ avoided (g/KWh)	733.02	CO$_2$ value (€/g CO$_2$)	0.00004

Air quality: pollutants removal (direct)

NO$_2$ removal (g/m^2/y)	2.33	NO$_2$ value (€/g)	0.0063
SO$_2$ removal (g/m^2/y)	1.98	SO$_2$ value (€/g)	0.0039
O$_3$ removal (g/m^2/y)	4.49	O$_3$ value (€/g)	0.0063
PM removal (g/m^2/y)	0.65	PM value (€/g)	0.0054

Carbon sequestration (direct)

C sequestred (g C/m^2/y)	164.97	C value (€/g CO$_2$)	0.00004
C to CO$_2$ (g CO$_2$/g C)	3.67		

Increment of roof longevity (direct)

	Investment avoided (€/20y/roof)	108
	Total Benefit Green Roof (€/m^2/year)	**2.91**

Rainwater barrels

The main co-benefit obtained from applying rain barrels is water savings. In this case, where water is obtained from seawater desalination, the reduction of water production implies energy savings. Moreover, since energy production is oil based, energy savings cause a reduction on carbon emissions and air pollution. Another benefit from using rainwater barrels is freedom from water restrictions. This is the economic value of avoiding drought related impacts, such as loss of amenity and other lifestyle benefits (CRC for Water Sensitive Cities, 2016). Figure 4.5 summarises all the processes considered to calculate these benefits.

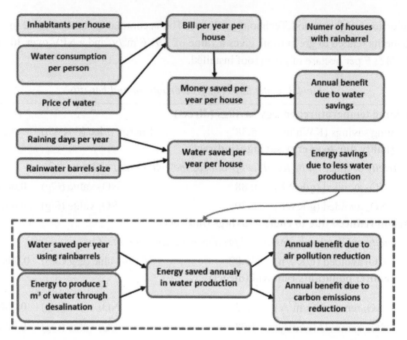

Figure 4.5. Benefits calculation due to rainwater barrels installation

Considering a population of 40,009 inhabitants and annual water production of $4836 \times 103 m^3$ (Central Bureau of Statistics, 2009), the average consumption of water is $10 m^3$ per month. Assuming three people per house (Department of Statistics, 2011) and an average water price of $3€/m^3$, the cost of water per house and per month is 89€. If a barrel of 600 L is installed, and considering that in average it rains 12 days per month (Meteorological Department St. Maarten, 2018), about $7.25 m^3$ are saved per house per month. It means that 22 €/house/year are saved.

Drinkable water is obtained in the island through seawater desalination, in this process energy is consumed and CO_2 released. Therefore, savings on water consumption are indirect savings of energy and less CO_2 released. Energy consumption due to seawater desalination is estimated as 3 to 4 kWh/m^3 of water produced (Elimelech and Phillip, 2011). Considering the same energy price than in the case of green roofs, 6.5€ are saved per house and per month. Also, like in the case of green roofs, oil based energy production releases air pollutants, following a similar calculation than for the case of green roofs, the benefit in this case is 3.9€/house/year. Regarding the decrease of CO_2 released, it is estimated that 1.4 to 1.8 Kg of CO_2 are released by each m^3 of desalinated water (Elimelech and Phillip, 2011). Again, following a similar calculation than in the case of green roofs, the benefits due to carbon emissions reduction is 5.3 €/house/year.

Finally, for each house free of water restrictions, the benefit is equivalent to the willingness to pay for the impact of droughts avoided, estimated as 74 to 137 €/house/year (CRC for Water Sensitive Cities, 2016). In this work the benefit considered is 74 €/house/year. The average roofs size is assumed as 150m².

The total benefit due to rainwater barrels installation is 2.82€/m² of roof connected to a rainwater barrel. With a NPV over 30 years of 433621 € per hectar of roof connected to barrels.

Pervious pavements

The main benefit considered due to pervious pavements installation is heat stress reduction. Cooler pavements reduce outdoor temperatures, decreasing the use of air conditioning, hence reducing energy consumption and the emission of CO_2 and air pollutants. Santamouris (2014) indicates that the consumption of energy for cooling is increasing de to urban heat insland effect. Moreover, according to Akbari et al. (2001), the demand of electricity in cities is increased by 2–4% per each °C of outdoor temperature. The installation of pervious pavements can reduce surface temperatures up to 4 °C, due to lower reflection and evaporation (Foster et al., 2011). Furthermore, since pervious pavements are more effective on reducing outdoor temperatures when are wet, better results are obtained in warm and humid climates (Santamouris, 2013). The results obtained in this work show a reduction of 12% of energy consumption in impacted houses. This number is in agreement with Santamouris (2014), who estimates an energy consumption increment of 13% due to urban heat island effect.

Energy savings due to pervious pavements installation were calculated considering that the impact of outdoors temperature reduction reaches the houses located directly in front of these pavements. Therefore, to estimate the number of houses impacted, the length of pervious pavements installed was taken into account as well as an area covering about 50m each side of the pavement. A 3% of energy reduction per each °C of temperature decreased was assumed. The temperature reduction achieved was assumed as 4 °C per each raining day, with a total of 145 raining days per year (Meteorological Department St. Maarten, 2018).

The average domestic electricity bill in Sint Maarten is 172€ per month and per house (Radjouki and Hooft Graafland, 2014). Therefore, the benefit due to energy savings is around 98€ per impacted house and per year. Regarding the indirect benefits of energy savings, the benefit due to air pollutants reduction is 12.8 €/house/year, while the benefit due to carbon emissions reduction is 22.6 € per year and per impacted house. Concerning the benefit of having cooled suburbs in summer, the value of this benefit is sustained in the willingness to pay for improvements in human thermal comfort and avoided halth care costs. This value is established between 30 and 51€/year per house which experiments a reduction of peak summer temperature of 2 °C (CRC for Water Sensitive

71

Cities, 2016). In this work an average value of 40€/year per impacted house is considered. The total benefit due to pervious pavements installation is 2.87€ per m^2 of pavement per year, with a NPV over 30 years of 440700 € per hectar of pavement.

4.3.5 Costs calculation

Cost values were taken from local and regional data, combined with inputs from literature review. The construction cost of pervious pavements considers a layer of permeable asphalt or concrete above a highly permeable layer of gravel. Taking also into account excavation, underdrain construction and contingencies. Maintenance is mainly cleaning since this measure is susceptible to clogging (Narayanan and Pitt, 2006).

Regarding the cost of rain barrels, we considered the cost of 600lts barrels and the pumping system. The operation cost in this case is the cost of energy needed for pumping. The cost was then calculated per square meter of roof apporting to the barrel. In the case of green roofs, materials and installation of extensive green roof were considered for investment cost calculation, while maintenance includes mainly inspection, vegetation care and roof reparations (Narayanan and Pitt, 2006). To estimate the capital costs of earthen open detention basins, regression equations calculated by Narayanan and Pitt (2006) were used. This calculation considers soil movement and compactation assuming that the soil needed is available, and there is not rock excavation nor groundwater problems.

Finally, to calculate the cost of pipes, lookup tables for reinforced concrete pipes (Narayanan and Pitt, 2006) were considered together with excavation, bedding and backfill costs. To estimate operation and maintenance costs an annual value of 3% of capital cost was included. A summary of results for the costs previously described is presented in Table 4.4. The costs are in euros and actualized according to consumer price index (see Equation 4.2).

Table 4.4. Total costs presented as net present value over 30 years

Green-Blue infrastructure	Total Present Value$_{30}$ (€/m^2)
Pervious pavements	161
Rain barrels	20
Green roofs	278
Open detention basin	349
Grey infrastructure (pipes diameter)	**Total Present Value$_{30}$ (€/m)**
800 mm	719
1000 mm	894
1500 mm	1534
2000 mm	2947
2500 mm	3615

4.3.6 Strategies development and results comparison

The evaluation is performed for different alternatives, considering each measure applied separately and combinations of them. The areas considered for each infrastructure are the maximum possible application for each case. In the case of pervious pavements, we assumed that these pavements can be installed on 50% of roads in zones of low slope. As a result, this measure covers a maximum of about 5% of the total area. In the case of green roofs and rainwater harvesting, the assumption was that these two measures cover the total area of roofs which represents about 15% of the total area. In the case of detention basins, 12 structures were considered to storage runoff from steep areas, with a total volume of 90,750m^2. Regarding grey infrastructure, a 2500mm pipe was applied.

If the ratio between primary benefit and total cost is plotted for each strategy (in blue in Figure 4.6), we observe that the application of pipes (Pi) appears as the best strategy. This option offers benefits more than two times higher than costs. However, when primary benefits and co-benefits are presented together as total benefits, and the ratio of this value vs. cost is analysed (in red in Figure 4.6), other strategies appear as good options too. In this case, the options of rainwater harvesting (RH) and its combinations with open detention basins (ODB) and Pi, also offer benefits higher than costs. In particular, Pi, RH and the combination of both of them (Pi + RH) appear as the most promising strategies from this analysis.

With the objective of analysing further these options, total cost, damages reduction, total benefits and residual damages are plotted in Figure 4.7. Here we can see that the option of RH, which presented the best ratio between total benefits and cost in the previous analysis, does not perform well for the primary benefit, since it presents high residual damages. The strategy with lowest residual damages is the combination of ODB, Pi and RH. Another option, with similar residual damages but lower costs, is the combination of OBD and Pi. But this option does not offer co-benefits. The rest of the alternatives present residual losses higher than avoided losses.

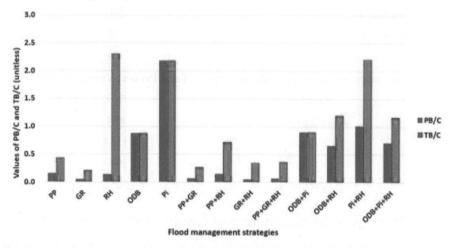

Figure 4.6. Primary benefit vs. cost (PB/C) and total benefits vs. costs (TB/C) for each strategy

There are some case specific characteristics that may explain the low effectiveness of most of the alternatives. Due to local topography, measures cannot be placed in steep areas. Runoff from these areas is managed using open storages, while green and grey measures are placed in the low-lying flat areas. This arrangement can result on a lower efficacy of green and grey measures. Another factor that may have influenced these results is the selection of measures through a multi criteria analysis. Perhaps there is a better combination of green measures to reduce runoff, which could have diminished the advantage of the blue and the grey options. A third and final factor that may be affecting these results are the limitations for grey infrastructure design because of the existent drainage system configuration. Maybe another design could have result on a more efficient grey option compared with the blue one. In any case, two consequences are observed. First, green measures appear as considerably less efficient than the grey and blue ones. Second, the importance of co-benefits to achieve higher benefits than costs is more evident. Despite that, the results obtained are in general as expected if we compare with previous studies (Ossa-Moreno *et al.*, 2017; Urrestarazu Vincent *et al.*, 2017; Wild *et al.*, 2017). Grey and blue infrastructures are cheaper than green measures, so it is expected a higher efficiency from them. Besides, a combination of green, blue and grey measures appears as the best alternative.

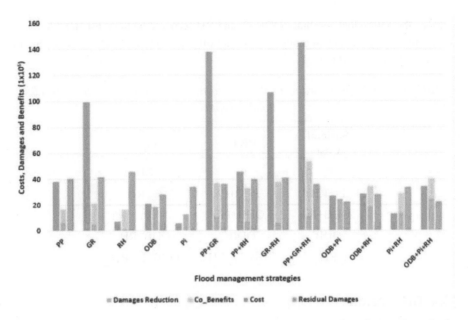

Figure 4.7. Total costs, total benefits (damages reduction + co-benefits) and residual damages for each strategy

Klijn *et al.* (2015) studied how decision-making changes when different criteria are considered. Following a similar analysis, three different rankings were developed (Table 4.5). The first ranking represents the minimisation of societal cost, calculated as the addition of cost (implementation plus maintenance) and residual damage. The second ranking maximises the ratio between primary benefit and cost, this criterion represents the traditional approach, in which the only benefit taken into account is damages reduction. The third ranking maximizes the ratio between total benefits and cost, with total benefits being the sum of damages reduction plus co-benefits. If only the first and second rankings are considered (traditional approach), pipes (Pi) is the strategy to apply in this case. However, if the first and third rankings are considered, also rain barrels (RH) appear as a good option. Besides, the combination of Pi and RH, which is the second option in both cases, seems to be the strategy to choose. Moreover, looking at the third and fourth positions in these rankings, open detentions (ODB) should be included in further analysis as well. This is, combinations of green, blue and grey measures should be evaluated in this case.

In this work only material damages are considered, including buildings, transport and infrastructure. However, if other material damages not considered here, or non-material damages (such as physiological trauma or risk of life) are considered, the importance of reducing residual damages turns crucial. Besides, open detention basins can offer co-benefits such as recreational spaces, which are not considered in this work. For these reasons the option that combines ODB, Pi and RH seems to be the most suitable for the case under study here, even though it does not offer the lowest societal cost, neither the best ratios PB/C or TB/C.

Table 4.5. Rankings according to: societal costs, PB/C (primary benefit/cost), TB/C (total benefit/cost)

Societal Cost	PB/C	TB/C
Pi	Pi	RH
Pi+RH	Pi+RH	Pi+RH
ODB+Pi	ODB+Pi	Pi
ODB	ODB	ODB+RH
RH	ODB+Pi+RH	ODB+Pi+RH

4.4 DISCUSSION

The main goal of this work was to see how economic viability of different flood management strategies changes when co-benefits are included into the analysis. To achieve this, we presented monetary values of flood damages reduction, co-benefits and costs for different combinations of green-blue-grey infrastructures considering several rainfall intensities. The analysis of results shows how the selection of strategies changes when cobenefits are considered. Only grey measures appear as feasible if co-benefits are not taken into account. However, when these secondary benefits are included, combinations of green-grey measures and green-blue-grey measures appear as economically viable and, at the same time, good to ensure the primary benefit of flood risk reduction. Similar results have been obtained previously by Ossa-Moreno *et al.* (2017) and Engström *et al.* (2018).

Our analysis is based on literature review and local and regional data for the case studied here. The numerical results obtained, and rankings of measures developed are valid under the assumptions made for this case. This study does not attempt to provide precise cost and benefits data. Rather, the objective of this work was to show how a holistic approach can help to choose sustainable solutions for urban flood mitigation. Moreover, the analysis presents uncertainties and constraints based on data availability and particularities about local data and local issues.

Despite these uncertainties and constraints, similar values of costs and co-benefits have been found in other works. Comparable values of costs and lifespans for green roofs and pervious pavements have been presented by Engström et al. (2018) and Foster et al. (2011). Regarding co-benefits, Foster et al. (2011) mention values between 25 and 50%

of reduction in water consumption due to water barrels installation, we obtained 23% of water savings in this case. The same work presents examples with values between 15 and 45% of annual energy savings due to green roofs installation, while they obtained 4% of energy reduction because of lower cooling costs. In this work we obtained annual energy savings of 12% due to green roofs installation. Finally, we assumed 4 °C of surface temperature reduction due to pervious pavements application and their impact on heat stress mitigation, again, a similar value than the one obtained by Foster et al. (2011). We obtained 12% of energy savings in houses located close to pervious pavements, a similar result was obtained by Santamouris (2014).

In this work, only case relevant and easily quantifiable co-benefits were considered. As a result, the values of co-benefits obtained were relatively low. These values can be enlarged if other benefits are considered, such as enhancement of biodiversity, groundwater recharge, water quality, property value, etc. Future work should include a broader range of co-benefits.

Concerning runoff reduction, the results obtained indicate low values of damages reduction when only green measures were applied. This result is adequate considering that five return period rainfalls were used for the analysis, including extreme events with 50 and 100 years return period. Other authors have found green and blue-green measures effective on providing flood reduction benefits (Kong et al., 2017; Haghighatafshar et al., 2018), but the reduction of this effectiveness under strong rainfall events has been argued as well (Zölch et al., 2017; Versini et al., 2018).

For the case studied here, rainwater harvesting appears as an efficient measure if co-benefits are taken into account. However, this measure is not effective for flood management. This remarks the importance of keeping the focus on the primary function for which the measures are applied. If the focus is shifted to co-benefits maximisation, there is the risk of not achieving the flood mitigation pursued. The option with best performance regarding flood damages reduction is the combination of the three measures: rainwater harvesting, open detention basins and pipes. While the second highest value of flood damages reduction is obtained combining only open detention basins and pipes.

Following a traditional approach, the pipes system appears as the only measure which achieves benefits higher than costs and effective flood damages reduction. However, through this work we proved that net benefits can be much enlarged if pipes are combined with rainwater harvesting. As a result, the combination of these two measures is the best option if the objective is to maximise benefits in the case here studied. But, if the main objective is to minimise flood damages and maximise positive net benefits, the best option will be a combination of pipes with rainwater harvesting and open detention basins.

Future work is needed to further understand how the flood mitigation and co-benefits can be maximised while costs are minimised. Urban drainage systems are complex systems and the combination of different strategies to achieve several different benefits makes the problem even more complex. Additional work should focus on understanding how different combinations of measures can be better designed to improve efficiency.

4.5 CONCLUSIONS

In this chapter a method to include the monetary analysis of cobenefits into a cost-benefits analysis of flood risk mitigation measures was presented. Traditionally, the benefits of green infrastructure other than flood risk reduction were not taken into account in decision making processes. Several such co-benefits were considered in this work: water savings, energy savings due to less cooling usage, air quality improvement and carbon sequestration. The above approach was then applied to a case study, comparing costs and benefits with and without co-benefits. Different intervention strategies were considered, using green, blue and grey measures.

The results obtained illustrate in quantitative terms how the viability of green and blue infrastructure for flood mitigation can be improved substantially when co-benefits are considered. In the case analysed here, costs outweigh benefits for all the green strategies if co-benefits are not included. This means that the traditional grey option appears as the only economically viable strategy if co-benefits are not considered. Thus, it is important to assess co-benefits when identifying best adaptation strategies to improve urban flood risk management, otherwise green infrastructure is likely to appear less efficient than conventional grey infrastructure.

When co-benefits are considered for the same case, the option of rainwater harvesting offers benefits higher than costs. However, this alternative has bad performance achieving the primary benefit (flood damages reduction). This issue can be solved, and net benefits maximised, if rainwater harvesting is combined with pipes, i.e. if green-grey strategy is used. Moreover, the reduction of flood damages is maximum, maintaining positive net benefits, if these two measures are mixed with open detention basins (green-blue-grey strategy). Consequently, a mix of green, blue and grey infrastructures is likely to result in the best adaptation strategy as these three tend to complement each other. Grey (and blue) infrastructure excels at reducing the risk of flooding whilst green infrastructure brings in multiple additional benefits that grey infrastructure cannot deliver.

5

EXPLORING TRADE-OFFS AMONG THE MULTIPLE BENEFITS OF GREEN-BLUE-GREY INFRASTRUCTURE

Climate change is presenting one of the main challenges to our planet. In parallel, all regions of the world are projected to urbanise further. Consequently, sustainable development challenges will be increasingly concentrated in cities. A resulting impact is the increment of expected urban flood risk in many areas around the globe. Adaptation to climate change is an opportunity to improve urban conditions through the implementation of green-blue infrastructures, which provide multiple benefits besides flood mitigation. However, this is not an easy task since urban drainage systems are complex structures. This chapter focuses on a method to analyse the trade-offs when different benefits are pursued in stormwater infrastructure planning. A hydrodynamic model was coupled with an evolutionary optimisation algorithm to evaluate different green-blue-grey measures combinations. This evaluation includes flood mitigation as well as the enhancement of co-benefits. We confirmed optimisation as a helpful decision-making tool to visualise trade-offs among flood management strategies. Our results show that considering co-benefits enhancement as an objective boosts the selection of green-blue infrastructure. However, flood mitigation effectiveness can be diminished when extra benefits are pursued. Finally, we proved that combining green-blue-grey measures is particularly important in urban spaces when several benefits are considered simultaneously.

Based on Alves A., Vojinovic Z., Kapelan Z, Sanchez A., Gersonius B. (2019). "Exploring trade-offs among the multiple benefits of green-blue-grey infrastructure for urban flood mitigation", Science of The Total Environment, In Press, doi.org/10.1016/j.scitotenv.2019.134980.

5.1 INTRODUCTION

Population growth and climate change effects present a growing challenge in urban spaces (United Nations, 2014; EEA, 2016; Kabisch *et al.*, 2017a). In particular, water managers will have to deal with more frequent extreme weather events, such as higher rainfall intensities which will increase urban flooding and water pollution (IPCC, 2012; Jha et al., 2012). Additionally, other problems are expected to deepen in urban spaces around the globe due to these changes, for instance heat waves, droughts and air pollution (EEA, 2016). Consequently, the consideration of multiple benefits during urban infrastructure planning is important in order to develop sustainable solutions, which can help cities to be more resilient to worsening future conditions (Lundy and Wade, 2011; IPCC, 2012).

Adaptation to climate change can be seen as an opportunity to improve urban conditions through the implementation of green-blue infrastructures which have the capacity of providing multiple benefits (EEA, 2016; Kabisch et al., 2016). Moreover, according to Elmqvist et al., (2015) investments in enhancing green infrastructure in cities are ecologically and socially required, but also economically viable. This qualities can be assessed through the acknowledgement and quantification of the benefits provided by these infrastructures. Such information is a crucial input for decision-makers.

Urban spaces represent complex systems, since natural, social and built environments interact. Furthermore, drainage systems are also complex structures, which can be integrate many different measures, imply significant investments and high uncertainties regarding future conditions (Jha et al., 2012; Simonovic, 2012). Green-blue infrastructures offer a holistic perspective to build resilience and address complex urban challenges, in which several problems need to be addressed at the same time, with limited resources and space constraints (Vojinovic, 2015; Frantzeskaki et al., 2019).

Urban drainage terminology has expanded in the last decades, consequently similar concepts are named with different terms. For instance, BMPs (best management practices), LIDs (low impact development), WSUD (water sensitive urban design), SuDS (sustainable drainage systems), GBI (green-blue infrastructure), EbA (ecosystem-based adaptation) and NBS (nature-based solutions) are largely used (Fletcher et al., 2014). Green infrastructure is defined as a network of multifunctional green spaces which maintain and enhance ecosystem services and resilience (Tzoulas et al., 2007; Naumann et al., 2011; European Commission, 2012a). In this work, the term green-blue infrastructure is used, referring to the concept of measures or solutions based in nature or natural processes.

While traditional drainage systems depend on grey solutions, resilience against future environmental threats cannot be achieved with these approach alone (Browder et al.,

2019). Besides, even though GBI has proved to be effective reducing flood risk (Kong et al., 2017; Haghighatafshar et al., 2018; Versini et al., 2018) and can contribute to multiple benefits, this might not be enough to cope with extreme future climate hazards (European Commission, 2012a; Demuzere *et al.*, 2014; Kabisch *et al.*, 2017a). Consequently, new tendencies suggest that the combination of green-blue and grey infrastructure may offer a novel generation of solutions to enhance community's protection (Browder et al., 2019). According to Frantzeskaki et al. (2019), green infrastructures should be complemented with technology-based solutions, hence more research is needed on how to combine multiple solutions to maximize climate adaptation in cities.

Despite much research has been done showing the advantages of using GBI, traditional grey infrastructure continues to be widely preferred in urban areas throughout the world (Dhakal and Chevalier, 2017). Several barriers for GBI acceptance are identified, which comprise socio-political, institutional and technical barriers (O'Donnell *et al.*, 2017). From a technological point of view, while traditional approaches count with enough technical support and tools for decision making, GBI for stormwater management lacks sufficient technical references, standards and guidelines (Qiao et al., 2018). In particular, this support is lacking regarding the evaluation and quantification of additional benefits (IPCC, 2012). Another commonly identified barrier is uncertainty about long-term performance and cost-effectiveness compared to conventional solutions (Davis et al., 2015). Therefore, further actions are needed to increase the acceptance of GBI over grey infrastructure for water management. To achieve this, the emphasis on the provision of multiple benefits in addition to flood protection is a key element (Kabisch *et al.*, 2017a).

Several works focus on the selection of GBI considering co-benefits and stakeholders' involvement (Alves *et al.*, 2018b; Miller and Montalto, 2019; Santoro *et al.*, 2019) . However, more quantitative results regarding the impacts of these measures on flood mitigation and co-benefits enhancement are needed (Pagano *et al.*, 2019). Regarding this, hydrodynamic models are widely used to select and design flood risk management strategies (Teng et al., 2017). But, the problems to be solved are usually complex and can have many possible solutions. In these cases is when optimisation evolutionary algorithms become helpful since they can be linked to hydrodynamic models to explore large solutions spaces, allowing the evaluation of many more options and trade-offs (Maier et al., 2019). Even though evolutionary optimisation processes imply high computational efforts, these algorithms offer a very useful tool for helping decision-making in complex systems, and in particular in the case of water resources management (Nicklow et al., 2010; Maier et al., 2014).

Previous research have shown that optimisation algorithms are a valuable tool to help solving stormwater management problems (Delelegn *et al.*, 2011; Vojinovic *et al.*, 2014; Woodward *et al.*, 2014). Besides, some works have included green-blue infrastructure into these frameworks (Zhang *et al.*, 2013; Alves *et al.*, 2016b; Behroozi *et al.*, 2018).

However, few works included the attainment of co-benefits from green-blue infrastructure as an extra objective when trying to solve stormwater related problems (Urrestarazu Vincent et al., 2017; Di Matteo et al., 2019). Furthermore, even though trade-offs when targeting multiple benefits have been considered in the past (Demuzere et al., 2014; Hoang et al., 2018), none of these works perform a quantitative analysis of these trade-offs. In addition, to the best of our knowledge not previous work focuses on compromises between primary and secondary benefits when comparing among green-blue and grey infrastructure application.

In response to these limitations, this work focuses on a method to quantitatively analyse the trade-offs when different benefits are pursued in stormwater infrastructure planning. First, different green, blue and grey measures and their combinations are considered in the evaluation of their performance to achieve flood risk reduction. Second, we include into the performance analysis the achievement of other benefits. Then, we investigate how the effectiveness of solutions regarding the primary function of flood risk reduction varies when the extra benefits are added. Finally, the changes in the composition of optimal solutions when the pursued objective is switched are analysed. In other words, we analyse how green, blue and grey measures are selected in different cases.

5.2 METHODOLOGICAL APPROACH

5.2.1 Strategies selection, cost and co-benefits calculation

The optimisation of urban drainage strategies is a complex and time-consuming analysis. Therefore, the reduction of alternatives to be analysed is an important step. A pre-processing method is applied to choose among drainage measures (see Figure 5.1a). Through this step, the number of options is reduced before starting the optimisation process. In this case we use a multi-criteria analysis in which local characteristics and needs are considered. This method is based on questions answered by local stakeholders (see Appendix A). The questions are about flood characteristics and local physical conditions, which are inputs for measures screening. In addition, the stakeholder selects weights establishing which are the preferred co-benefits in the area. The final step consists on defining the order of importance among flood mitigation, costs minimisation and co-benefits enhancement. Then, the answers are processed following the multi-criteria procedure. The result is a ranking of applicable measures for the area, more details can be found in Alves *et al.* (2018a).

An important aspect pursued with the use of this multi-criteria selection method is to improve the stakeholders' acceptance of the measures selected. By taking into account local preferences and necessities when choosing among options, and considering the opinion of local stakeholders from the very beginning, it is expected that the final solution will be better accepted for implementation (Kabisch *et al.*, 2017a; Bissonnette *et al.*, 2018). Moreover, this multi-criteria method can be used with diverse stakeholders, allowing to take into account their different objectives.

The next step after the identification of applicable measures is the development of possible combinations of green-blue and grey measures. These combinations are called here strategies and are selected after performing a spatial analysis of the study case. For instance, open detention basins are chosen if there is availability of open public spaces where to locate them, and green roofs are chosen if there exist adequate roofs where to build them. Afterwards, these strategies are evaluated quantitatively considering its flood risk reduction performance, co-benefits enhancement capacity and life cycle costs. To evaluate the selected strategies regarding co-benefits, we need first to identify direct and indirect co-benefits provided by each measure.

Several previous studies help us to recognise the multiple benefits delivered by GBI, see for example Woods-Ballard et al. (2007), Center for Neighborhood Technology (2010) and Horton et al. (2016). These works also offer quantitative data about the benefits, which allow us to calculate the annual values of those co-benefits which can be directly monetised (Alves et al., 2019). For example, water saving from rainwater barrels installation provides the co-benefit value of reducing the water bill accumulated along the year. The present value of these co-benefits is then calculated defining the measure's lifetime and a discount rate. These values will be given per unit of measure and will be an input into the optimisation process.

The aim of this study is to compare among green-blue, grey and hybrid strategies for flood mitigation from an economic point of view, and show how this comparison changes when co-benefits are considered. There are several co-benefits not easily quantifiable in economic terms, such as aesthetic value and biodiversity enhancement. Even though these co-benefits could be an important driver for decision making, they are not considered here because are not representable in a cost-benefits analysis.

Finally, to calculate the total cost for each measure local prices and literature review data are used (e.i. Narayanan and Pitt, 2006). Investment and annual operation and maintenance costs considered through the lifespan period of each infrastructure. Then the values are converted to the same year valuation using the consumer price index. Once more present values of these costs per unit of measure are calculated and will be an input to the optimisation process. More details about costs calculation are given in Alves et al. (2019).

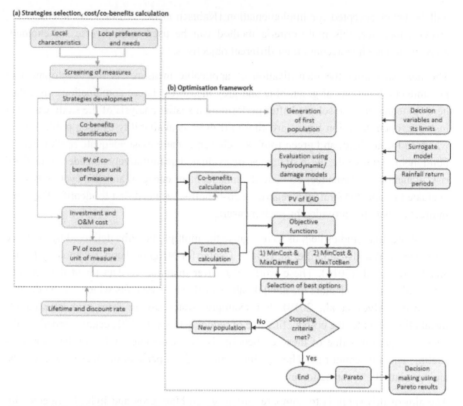

Figure 5.1. Methodological approach (a) Strategies selection and cost-benefits calculation, (b) Optimisation process; with PV: present value, O&M: operation and maintenance, EAD: expected annual damage, MinCost: minimum cost, MaxDamRed: maximum damage reduction, MaxTotBen: maximum total benefits

5.2.2 Optimisation framework

According to Maier et al. (2019), in a traditional or informal process the selection of solutions is based on experience or intuition. In the case of flood management this would represent the type of measures, its size and where to locate them. Then, the performance of selected solutions is evaluated using for instance a hydrodynamic model. Afterwards, other options would be evaluated with the aim of improving performance. However, when many decision variables exist it is unlikely to find even a near optimal solution. The authors argue that is in these cases that formal optimisation helps to identify optimal solutions in an efficient manner.

The multi-objective optimisation process followed in this work is presented in Figure 5.1b. The first step is problem formulation, this includes the establishment of decision variables, its search boundary values, and objective functions for the problem under analysis. In this case, the decision variables are the areas covered by the different drainage measures applied. The optimisation process will evaluate different options, each one with different measure's application areas. The definition of the minimum and maximum value of the areas is based on land use analysis. This is done measuring the surface covered by roofs, pavements, and open spaces with the use of aerial images and GIS analysis. Using this analysis we can define maximum values for each variable. For instance, a maximum of 50% of pavements with less than 5% slope covered by pervious pavements, or a maximum of 75% of roofs connected to rainwater barrels.

Concerning objective functions, we defined three objectives: total cost minimization, maximization of flood damage risk reduction, and maximization of total benefits:

$$O_1 = Min \left\{ \sum_{x=1}^{N} \left[\left(C_{Inv\_x} + \sum_{y=1}^{LT} \frac{C_{O\&M\_x}}{(1 + i/100)^y} \right) * \sum_{j=1}^{SC} S_{xj} \right] \right\} \tag{5.1}$$

where C_{Inv-x} is the investment cost for the measure x, $C_{O\&M-x}$ is the operation and maintenance cost of the measure x, LT is the lifetime considered for the measures, i is the discount rate, and S_{xj} is the application size of the measure x in the sub catchment j.

$$O_{2.1} = Max \left\{ EAD_{Max} \right.$$

$$\left. - \sum_{y=1}^{LT} \left[\Sigma_j \left(\left(\frac{TD_{RPj+1} + TD_{RPj}}{2} \right) * \left(\frac{1}{RP_j} - \frac{1}{RP_{j+1}} \right) \right) \middle/ \left(1 + \frac{i}{100} \right)^y \right] \right\} \tag{5.2}$$

where EAD_{Max} is the expected annual damage for the current situation which represents the maximum damage (before measures are applied), TD is total damage obtained from the model once the measures have been applied (includes residential, commercial, infrastructural and transport damage), RP is the rainfall return period, i is the discount rate, and LT is the lifetime considered for the measures.

$$O_{2.2} = Max \left\{ EAD_{Max} - \sum_{y=1}^{LT} \left[\Sigma_j \left(\left(\frac{TD_{RPj+1} + TD_{RPj}}{2} \right) * \left(\frac{1}{RP_j} - \frac{1}{RP_{j+1}} \right) \right) \middle/ \left(1 + \frac{i}{100} \right)^y \right] \right.$$
$$\left. + \sum_{x=1}^{N} \left[\left(\sum_{y=1}^{LT} \frac{Annual\ Co - Ben_x}{(1 + i/100)^y} \right) * \sum_{j=1}^{SC} S_{xj} \right] \right\} \tag{5.3}$$

where EAD_{Max} is the expected annual damage for the current situation which represents the maximum damage (before measures are applied), TD is total damage obtained from the model once the measures have been applied (includes residential, commercial, infrastructural and transport damage), RP is the rainfall return period, i is the discount rate, and LT is the lifetime considered for the measures, Annual Co-Ben$_x$ are the co-benefits obtained in one year from the measure x, and S_{xj} is the application size of the measure x in the sub catchment j.

Since all costs, co-benefits and flood damage are in monetary units, we could solve a single objective problem by maximizing net benefits (total benefits – costs). A single objective problem is much easier to solve than multiple objective ones, nevertheless in this work we optimize for two objectives separately. Even though computationally more demanding, this approach gives a detailed trade-offs picture between the objectives which would otherwise not be possible. This, in turn, helps decision makers to make better informed decisions at the end.

Concerning the experimental setup, the objective functions are used for options evaluation in two different cases, in which two objectives are pursued. First, the optimisation problem is formulated with minimisation of total costs and maximisation of flood damage reduction (O_1 and $O_{2.1}$) as objectives. The second optimisation problem is reformulated from the first one by changing the second objective to maximisation of total benefits (i.e. using O_1 and $O_{2.2}$ as objective functions). In the first objective function (O_1), the value to be minimised is total cost, which comprises investment and maintenance costs for the different drainage measures considered. The total cost is calculated multiplying the present value of cost per unit of measure, estimated in the past step, times the size of measures defined for each option during the optimisation process (see Equation 5.1).

The evaluation of options regarding flood damage reduction is performed using the hydrodynamic model EPA SWMM (Rossman, 2010). Using a 1D-1D model we estimate flood water depths at several locations in the area under different rainfalls. In this 1D-1D model, two parallel conduits connected among them are defined, one representing the drainage system and the other one representing the streets. Flooding occurs when water is accumulated in the conduit representing the streets. Then a surrogate model is used to estimate damages. The surrogate model links the 1D-1D model results with pre-calculated results from a 1D-2D model to estimate water depths and corresponding flooding damage values (see Figure 5.2), more details can be found in Alves et al. (2019). Through this method the total flooding damage can be calculated and it is possible to calculate the reduction of damage, which will be our primary benefit. Residential, commercial, infrastructure and transport damage are considered here. These damage values are used to calculate the risk of flooding as the expected annual damage (EAD) for different rainfall events (Delelegn et al., 2011). Then, we maximise the flood risk reduction (O_2) which is the difference between maximum EAD (without measures application) and the EAD obtained applying measures (Equation 5.2). This value is also used in the third objective function (O_3), in which total benefits are maximised. To achieve this we add co-benefits to the equation, which are the result of multiplying the present value of co-benefits per unit of measures, times the size of measures defined for each option during the optimisation process (Equation 5.3).

Once decision variables and objectives are established, the optimisation process follows the steps of the genetic algorithm NSGA-II applied in this work (Deb et al., 2002). The decision variables in this case are coded as GA chromosomes using integer values, these values represent the areas covered by the applied measures. In the first step, the optimisation process evaluates an initial random generation using the objective functions. Then the best options are selected and a new population is created applying concepts of crossover and mutation. This new population is then evaluated and the same process is repeated in a loop until the stopping criteria is met. The stopping criteria in this case is the number of generations to be analysed. There are other parameters which are also inputs for the optimisation process besides objectives and variables: population size, number of generations, crossover and mutation rates. These values were defined through a sensitivity analysis. Finally, when the stopping criteria is met, several "best options" are presented in a Pareto plot. The present optimisation framework builds upon and connects to previous work (Vojinovic et al., 2006; Vojinovic and Sanchez, 2008; Barreto et al., 2010).

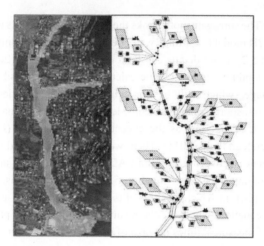

Figure 5.2. 1D-2D flood modelling result (left) and 1D-1D model representation (right)

5.3 RESULTS

5.3.1 Study area description

The study area is the catchment Cul De Sac, one of the most vulnerable areas to flooding in the Dutch side of Sint Maarten Island, located in the Caribbean region (see Figure 5.3a). This catchment has an area of 509Ha and the land use is predominantly residential, with some dispersed commercial areas in the lower part. Elevation ranges from near sea level to hilly areas with until 380m altitude at the northern borderline and the catchment is divided in 12 sub-catchments (see Figure 5.3b).

The catchment has several characteristics contributing to flood risk. For instance, urban areas are situated on low-lying zones (see Figure 5.3c). Besides, the existing drainage system which is composed by channels, has not enough capacity to avoid flooding (UNDP, 2012). In addition, most of the streets are narrow limiting the enlargement of these drainage channels (Vojinovic and van Teeffelen 2007). Recurrent inconveniences such as transport disruptions occur during small rainfall events. Whereas heavy rainfall causes large-scale flooding with damage to residential and commercial buildings (UNDP, 2012).

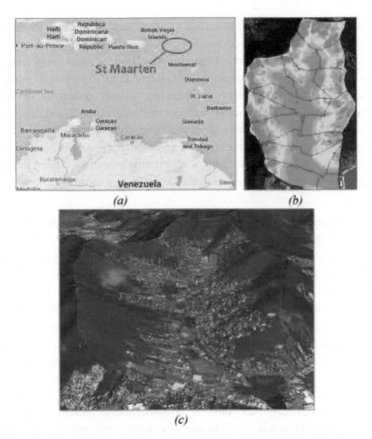

Figure 5.2. (a) Sint Maarten location, (b) Catchment topography and sub-catchments division, (c) Cul De Sac aerial visualisation

5.3.2 Screening of measures and cost-benefits calculation

A questionnaire was filled by technical and political decision makers related to water management in the island (see Appendix A). The questions were about flood type, physical site conditions, drainage system characteristics, land use and preferred co-benefits for the area under study. The answers where used to apply the multi-criteria analysis described in Section 5.2.1 and illustrated in Figure 5.1a.

Regarding local characteristics, this analysis allowed us to conclude that the main flooding problem in the area is pluvial flooding. Furthermore, the soil has medium permeability with deep water table and bedrock. The surface's slope is larger than 5%, the sewer system is separate but there is also illegal combined sewer system. The main land use is residential with medium to low density. The availability of public spaces is less than 25% and there is low space availability along roads and sidewalks. Finally, combined sewer overflows were identified as a problem in the area. Regarding local preferences and needs, several co-benefits were identified. The most important co-benefits identified for this area were liveability improvement (heat stress reduction and aesthetics enhancement), socio-cultural benefits (community engagement, recreation and educational spaces), water quality enhancement (runoff pollutants removal) and environmental benefits (groundwater recharge and water reuse, and species habitat creation). Besides, decision makers identified flood problems affecting buildings and generating significant damage in the area as occurring every two years. Furthermore, they recognised budget restrictions when investing on infrastructure for flood management. Lastly, they described the achievement of co-benefits as a medium to low importance objective.

Using this information and through the screening of measures we identified preferred infrastructures to be applied in the area. Details about the method to select these measures can be found in Alves et al. (2018). This screening process established daylighting water courses and open water channels as preferred options for this case. This is in accordance with the practice of maintaining and enlarging (when possible) the existent channels system, already recommended by the study performed by UNDP (2012). Besides, the analysis detected pipes as a preferred option. This measure can be applied to enhance conveyance capacity of the existing channels, since there is limited space to enlarge them. Another selected measure was open detention basins. This result confirmed previous outcomes from a study performed in this catchment in which open detention ponds were identified as an effective flood management alternative (UNDP, 2012). Additionally, the multi-criteria analysis identified rainwater disconnection as another option for runoff management. Several measures could be applied to achieve this, but rainwater barrels was a preferred alternative in this case since it allows the reuse of water, an expensive and scarce resource in the island. Finally, measures that allow the infiltration of runoff were recommended. Due to the low availability of public spaces, the infiltration option chosen for this case was pervious pavements, to be applied in low slope and low traffic roads. In summary, the measures selected in this study for further analysis are: closed pipes (Pi), open detention basins (ODB), rainwater barrels (RB) and pervious pavements (PP). These options, and its combinations, were further evaluated using hydrodynamic modelling. The assessment was performed considering the existing channels system working at its current capacity.

Six strategies, or measures combinations, were chosen for further analysis using the optimisation framework. The objective is the comparison among green-blue and grey measures and its combinations. These six strategies are: rainwater barrels with pervious pavements (RB+PP), the same two measures combined with open detention basins and combined with pipes (RB+PP+ODB and RB+PP+Pi), the four measures combined (RB+PP+ODB+Pi), open detention basins alone (ODB) and combined with pipes (ODB+Pi). The selection of these combinations was based on the intention of comparing green-blue and traditional (or grey) measures. RB and PP are green-blue measures providing co-benefits, while ODB and Pi are traditional measures which do not provide co-benefits. The selected combinations represent then examples of only green-blue measures (RB+PP), different combinations of green-blue and traditional measures (RB+PP+ODB, RB+PP+Pi and RB+PP+ODB+Pi), and alternatives with only traditional measures which do not provide co-benefits (ODB and ODB+Pi).

The next step was to identify the relevant co-benefits provided by the selected measures and their importance for the case here studied. Rainwater harvesting barrels allow the reduction of drinking water consumption. This benefit is important in this case because drinking water in the island is produced using reverse osmosis, an expensive and high energy consumption technology (Elimelech and Phillip, 2011). In addition, water production and its cost have risen notoriously in the last 10 years in the area (Centrale Bank Curaçao en Sint Maarten, 2017) and the area goes through water shortages during high consumption hours (European Commission, 2012b). Pervious pavements allow urban cooling by means of lower reflection and higher evaporation (Foster et al., 2011). The benefits obtained are energy savings and carbon dioxide and air pollutants reduction (USEPA, 2012). Temperature reduction is especially important in areas with tropical weather, where energy consumption can increase between 2 to 4% per each extra Celsius degree (Akbari et al., 2001; Santamouris, 2014). Other benefits obtained from pervious pavements installation are water quality enhancement due to runoff filtration and groundwater recharge, which were also considered here. Even though recreation and liveability enhancement can be considered as co-benefits for open detention basins, these are not easily converted into monetary values and hence were not considered in the present study.

Afterwards, implementation and operation and maintenance costs were calculated. Details about how these costs and benefits values were calculated are presented in Alves et al. (2019). Table 5.1 presents the results of costs and co-benefits for each one of the four measures selected. In the case of Pi, the cost results are presented in €/m and for each diameter to be considered in the optimisation process. The cost of ODB is given in €/m2, considering an average depth of 1.5m in order to reduce the variables in the hydrodynamic model. The values corresponding to RB and PP are presented as €/m3 and €/m2, respectively. Regarding co-benefits, only these two measures provide them and PP presents a higher value than RB.

The values of costs and benefits showed in Table 5.1 are present values over a lifetime of 30 years with a discount rate of 5% rate (International Monetary Fund, 2016). The period of 30 years is considered as maximum before the necessity of replacement for green infrastructure (Pezzaniti et al., 2009; USEPA, 2012; Al-rubaei et al., 2013; Yong et al., 2013).

Table 5.1. Cost and co-benefits values for each selected measure (RB: Rainwater Barrel, PP: Pervious Pavement, ODB: Open Detention Basin, Pi: Pipes)

Measure		Cost		Annual co-benefit	
RB		1040	€/m³	30	€/m³
PP		160	€/m²	86	€/m²
ODB		350	€/m²	0	€/m²
Pi (mm)	800	720	€/m	0	€/m
	1000	895	€/m	0	€/m
	1500	1530	€/m	0	€/m
	2000	2950	€/m	0	€/m
	2500	3615	€/m	0	€/m

5.3.3 Optimisation results

The decision variables used in the optimisation process were the size of application of each measure. In the cases of RB, PP and ODB, these are the measures' application areas in each one of the 12 sub-catchments included in the hydrodynamic model. The ranges in which the area of each measure varies for each sub-catchment were defined through a land use analysis performed using aerial images (Table 5.2). In the case of pipes, a single pipe was chosen to follow the main channel path from the mid area of the catchment until its discharge. The variables are the diameters of the four segments which cover the pipe's extension. Depending on the strategy and the number of measures combined, the optimisation framework has different numbers of variables (see Table 5.3).

Different parameters can be chosen when applying the NSGA-II algorithm, such as population size, number of generations, and mutation and crossover operators. Several runs of the framework were performed to assess convergence and to choose the values of these parameters. Three indicators were used for Pareto fronts evaluation: the number of non-dominated solutions obtained in the final Pareto compared to the given number of initial population, the extent or spread of Pareto fronts with respect to the objectives, and the average space among solutions. We analysed the sensitivity of optimisation results to the parameters. Since the theoretical value of mutation is the inverse of decision variables (Mala-Jetmarova et al., 2015), this analysis was applied for the cases of maximum and minimum number of variables. Changing values of population (between 80 and 400), generations (between 20 and 80), crossover (between 0.2 and 0.9) and mutation (between 0.01 and 0.08), the values of number of non-dominated solutions, extend of Pareto curve and average space among solutions were evaluated. As a result, values of 350 individuals for population, 70 generations, 0.9 for crossover and 0.021 for mutation were selected to apply the optimisation framework.

The optimisation framework was applied twice for each one of these six strategies. Firstly, the framework was applied using the objective functions of cost minimisation (Equation

5.4) and flood risk reduction maximisation. Secondly, the objective functions of cost minimisation and total benefits maximisation were used. Rainfalls with return periods of 5, 10, 20, 50 and 100 years and 2 hours duration (UNDP, 2012) were considered to calculate EAD in objective functions $O_{2.1}$ and $O_{2.2}$.

Table 5.2. Value ranges of decision variables: area of roof connected to rain barrels (roof to RB), area of pervious pavement (PP), area of open detention basin (ODB), and pipe's diameter (Pi_Diam)

Sub-catchment	roof to RB (ha)		PP (ha)		ODB (m²)		Pipe	Pi_Diam (mm)	
	Min	Max	Min	Max	Min	Max		Min	Max
1	0	3.4	0	1.5	0	3000	1	500	2500
2	0	1.9	0	0.8	0	4000	2	500	2500
3	0	3.0	0	1.3	0	3500	3	500	2500
4	0	6.1	0	2.7	0	4000	4	500	2500
5	0	2.4	0	1.1	0	6000			
6	0	4.8	0	2.1	0	4000			
7	0	7.8	0	3.5	0	5000			
8	0	2.6	0	1.2	0	8000			
9	0	4.9	0	2.2	0	5000			
10	0	3.2	0	1.4	0	7000			
11	0	6.1	0	2.7	0	5000			
12	0	7.5	0	3.3	0	6000			

Table 5.3. Number of decision variables for each strategy

Strategy	Decision variables
RB+PP	24
RB+PP+ODB	36
RB+PP+Pi	28
RB+PP+ODB+Pi	40
ODB+Pi	16
ODB	12

For this case the first objective function is:

$$O_1 = Min\{C_{RB} * \textstyle\sum_{i=1}^{12} A_{RB} + C_{PP} * \sum_{i=1}^{12} A_{PP} + C_{ODB} * \sum_{i=1}^{12} A_{ODB} + C_{Pi} * \sum_{i=1}^{4} L_{Pi}\} \quad (5.4)$$

where C_{RB}, C_{PP}, C_{ODB} and C_{Pi} are the present values over 30 years of total costs of rainwater barrels, pervious pavements, open detention basins and pipes respectively. A_{RB}, A_{PP} and A_{ODB} are the areas of measures for each one of the 12 sub catchments, and L_{Pi} is the length of each one of the 4 pipes proposed for this case.

Figure 5.4 (a) and (b) show the obtained Pareto results. ODB combined with Pi (green) and the combination of all the measures (yellow) are the best performing strategies for flood risk reduction (Figure 5.4.a). However, costs exceed benefits when the cost is higher than 24 million € (Pareto fronts under grey line in the plot, where the grey line represents cost equal benefits) and hence these strategies are not cost efficient. The only benefit in this case is the reduction of flood damage and has a maximum of around 24 million € before the strategies are no longer efficient. The maximum present value of expected annual damage over 30 years in the current situation (without measures) is 47.5 million €. Therefore, the maximum damage reduction achieved applying these strategies is about 50% of that value.

All strategies achieve benefits higher than costs if we analyse the results obtained from total benefits maximisation (Figure 5.4.b). Even the combination of RB and PP (light blue) shows efficient results in this case, in contrast with the case of damage reduction maximisation. The best strategy in this case is the combination of RB, PP and Pi (orange) when the cost is lower than 19 million €. For higher costs the strategy achieving best results is the combination of the four measures (RB+PP+ODB+Pi). However, from the results obtained in the case of damage reduction, we observe that after 8 million € of cost the strategy RB+PP+ODB+Pi performs much better than RB+PP+Pi on flood risk reduction. As a result, even if slightly higher total benefits are obtained in the case of RB+PP+Pi for costs lower than 19 million €, the decrease on flood risk reduction seems not worth. Consequently, the combination of the four measures appears to be the best option. The Pareto curve for this strategy presents a slope change around the cost of 20 million €, suggesting that a solution around this cost will be the best option in view of the benefits obtained from the investment. In that case damage reduction will be around 23 million € (48% of the maximum damage) and total benefit around 40 million € (twice the cost).

Although the strategies including RB and PP deliver other benefits besides flood damage reduction (e.g. water and energy savings), these co-benefits cannot be appreciated in the results presented in figure 5.4a. To visualise this, we added the value of these co-benefits to the Pareto fronts obtained in the case of only flood damage reduction as second objective. The original optimal values are represented by DR and the results including co-benefits by DR+Co_Ben in Figure 5.5. Moreover, the results presented in Figure 5.4b do not allow us to see the performance of the strategies on flood mitigation. To appreciate this, we subtracted the co-benefits from the Pareto fronts obtained maximising total benefits as second objective. The Pareto fronts are represented by TB and the results without co-benefits by TB-Co_Ben in Figure 5.5. This is presented only for the four strategies providing co-benefits: RB+PP (Fig. 5.5a), RB+PP+ODB (Fig. 5.5b), RB+PP+PI (Fig. 5.5c) and RB+PP+ODB+Pi (Fig. 5.5d).

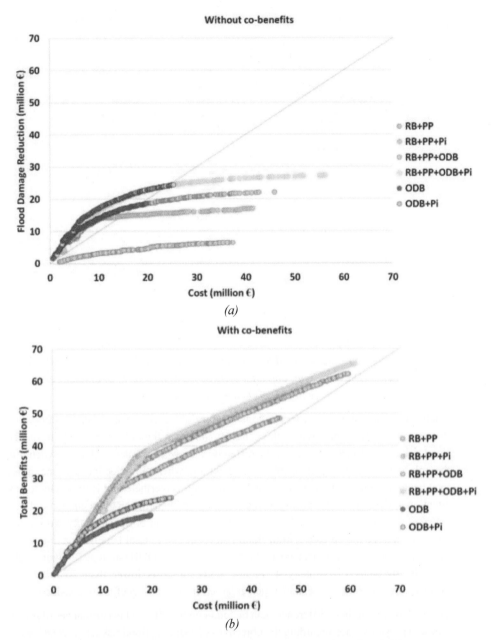

Figure 5.3. Pareto fronts obtained for the six strategies selected with (a) cost minimisation and flood risk reduction maximisation as objectives (b) cost minimisation and total benefits maximisation as objectives (grey line: costs = benefits).

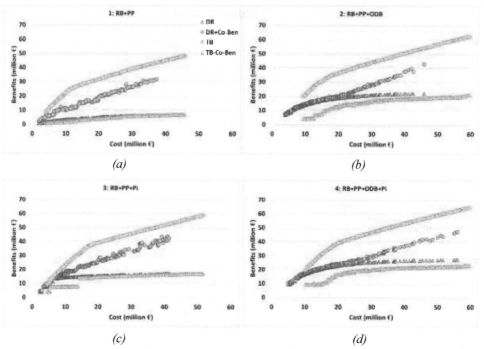

*Figure 5.4. Pareto fronts obtained for damage minimisation (DR) and the result adding
co-benefits (DR+Co_Ben), and total benefits maximisation (TB) and the result
removing co-benefits (TB-Co_Ben).*

Analysing these results, we observe a considerable difference between total benefits when
it is an optimisation objective (yellow circles) and when the objective is only to reduce
flood risk (blue circles). However, the differences between damage reduction when it is
the only optimisation objective (blue triangles) and when the objective is to maximise
total benefits (yellow triangles) is not that significant. Nevertheless, it is important to pay
attention to the impact of focusing on maximising total benefits on the reduction of flood
damage. In some cases, the reduction of flood damage can be substantially diminished
when we change the objective from flood risk reduction to total benefits maximisation.
This can be observed, for instance, in the cases of RB+PP+ODB (strategy 2, Figure 5.5b)
and RB+PP+ODB+Pi (strategy 4, Figure 5.5d) for costs lower than 20 million €.
Furthermore, this tendency can be much enlarged if more co-benefits are considered.

The explanation of these differences can be found on the different performances of green-
blue and grey measures regarding the objectives of reducing flood risk and increasing co-
benefits. It is expected that the optimisation algorithm will choose differently among the
measures, according to the pursued objective. To better understand this, an analysis of the
measures selected for optimal solutions in each case was performed. The analysis shows
the application value selected for each measure as a percentage of the maximum

measure's area that can be applied in each case (presented in Table 5.2). The results are shown in Figure 6, with damage reduction as objective in Figure 5.6a, and with total benefit maximisation as objective in Figure 5.6b.

Based on this analysis we can observe that RB and PP (blue and green in Figure 5.6a) are not preferred when the pursued benefit is to reduce flood risk. In this case, ODB (yellow) is the most applied option. However, when the sought benefit shifts to total benefits maximisation (Figure 5.6b), the application of RB increases sharply from a mean value below 20% to approximately 90% in all cases. Unlike ODB, RB is not a very effective measure for coping with runoff excess (i.e. reducing flood damage) but it is a low cost measure which provides substantial water and energy savings (main co-benefits in this case). Note that the usage of PP also increases when the second objective is to maximise total benefits, although to a lesser extent. This is because of PP is more expensive than RB and some of the co-benefits it provides are not so profitable, namely groundwater recharge and water quality enhancement.

Regarding the use of ODB (yellow), we can observe an important application decrease for strategies 2 and 4 when the objective is swiched to total benefits maximisation (Figure 5.6b). This is expected because, despite being an effective flood reduction measure, we have not considered co-benefits for this measure which makes it less attractive to the optimisation algorithm. In addition, note that the application of Pi (red) increases for all strategies in the case of total benefits maximisation (Figure 5.6b) relative to the case of flood risk reduction maximisation (Figure 5.6a). The explanation of this can be linked to the lower application of PP (green) in case of strategy 3 and ODB (yellow) in case of strategy 4, which implies less runoff reduction. As a result, optimal solutions focus on the improvement of system's conveyance to keep flood damage low.

Finally, major differences can be observed in terms of optimal strategy 4 composition when the second objective is changed. The application of GBI increases considerably, with mean values increasing from approximately 10% to 95% and 35% for RB and PP respectively. Besides, the mean application of ODB reduces substantially, from above 60% to 35%, and the mean use of pipes increases from approximately 10% to more than 40%. These changes imply the achievement of higher co-benefits, but also a decrease in the efficiency of flood risk reduction. This is the result already observed for strategy 4 (Figure 5.5d), in which a significant growth of total benefits is observed, but also a decrease of efficiency regarding flood mitigation. These changes suggest that special attention should be paid to the selection of second optimisation objective when multi functionality of measures is pursued. Local priorities should be considered closely with stakeholders in order to define the importance of each objective. These needs can then be represented in the optimisation process, for example incorporating a suitable weight for each objective, or with a careful post-process to analyse these trade-offs and make a decision accordingly.

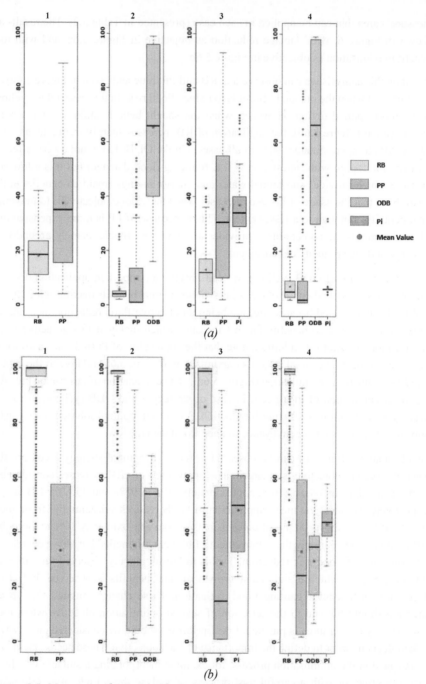

Figure 5.5. Measures selection analysis, with the objectives of cost minimisation and (a) flood risk reduction, (b) total benefits maximisation.

5.4 DISCUSSION

While the application of optimisation techniques in water resources enables the assessment of multiple options, it is often a time consuming task (Maier et al., 2014). The application of some form of pre-processing can shorten this time by reducing the number of optimisation options. This is even more important in cases with a bigger computational burden than the one here studied, for instance in cases with more extensive or complex drainage systems. However, the reduction of options needs to be done carefully not to lose useful information in the process and end up with sub-optimal solutions. In this work a systematic multi-criteria analysis was applied which allowed to shortlist measures and to interact with stakeholders, without losing information. The combination of this multi-criteria pre-process with a more quantitative post-process, which allows to compare strategies according to costs and benefits in the long term, is what is seen as novel in this research.

Besides, we have confirmed the usefulness of optimisation as a decision-making support tool in the context of stormwater management with green, blue and grey measures considered. The optimisation approach allows decision makers to identify the most effective solutions covering a wide range of costs and benefits. Moreover, they can visualise the effectiveness achieved for each level of investment, recognising which investment level gives them the highest return. The usefulness of optimisation methods for urban stormwater problems has been previously established, but co-benefits have been included into the analysis in few cases only (Urrestarazu Vincent et al., 2017; Di Matteo et al., 2019).

Since the simultaneous delivery of social, economic and environmental benefits by GBI increases the willingness to accept these solutions, awareness about these co-benefits is crucial to convince decision-makers about GBI implementation (EEA, 2012; Liu and Jensen, 2018; Qiao et al., 2018). Moreover, the economic analysis of these co-benefits can have a significant impact on decision-making by establishing evidence-based decisions and allowing its financial consequences to be visualised (EEA, 2016). The study presented in this work shows how the inclusion of co-benefits can encourage the selection of GBI for urban flood mitigation. Although the analysis presents constraints due to data availability and local characteristics, similar results concerning the effectiveness of this approach have been found in previous research (Elmqvist et al., 2015; Ossa-Moreno et al., 2017; Engström et al., 2018).

The inclusion of co-benefits in this analysis has been greatly limited due to the consideration of only those co-benefits easily represented in economic terms. Moreover, we chose only the most important co-benefits for this case, the ones having more economic relevance. Still, the results show how the inclusion of co-benefits analysis, even if limited, has an important impact encouraging the selection of GBI. A post analysis could be added to this framework to include a qualitative analysis of not monetisable co-benefits. Through this step, decision making could be further imporved considering the complete range of benefits achievable applying GBI and stimulating even more the selection of holistic and adaptive solutions.

Our results also highlight that combinations of green-blue-grey measures can be the best option for climate change adaptation, this result is compatible with other recent studies (WWAP/UN-Water, 2018; Browder et al., 2019). We proved that this is particularly important when several benefits are considered simultaneously. In urban spaces, where space is limited, the combination of green, blue and grey measures allows to maximise the efficiency with some measures performing best at flood risk reduction (open detention basins and pipes in this case) and other at co-benefits provision (rain barrels and pervious pavements in this case). Our results also state the importance of considering the achievement of co-benefits as a relevant objective from the beginning, when selecting and comparing among stormwater management options. When the focus is only on flood risk reduction, even if GBI is used, the co-benefits will be achieved as a side effect which can decrease largely its value.

The importance of considering trade-offs among objectives is also stressed in this work. This is particularly significant when adding new benefits while maintaining stormwater management as primary functions. Blue-green infrastructure can have low effectiveness decreasing flood damage in the case of high return period rainfalls (Zölch et al., 2017; Mei et al., 2018). Therefore, even if a strategy achieves the highest total benefit, attention has to be paid to the resulted compromise on flood damage reduction. A possible solution to this is to determine the importance of each benefit and add weights into the measures assessment framework. These weights will represent the level of trade-offs accepted and should be jointly defined with local stakeholders.

Finally, this work presents an analysis of which are the application values of measures selected in optimal solutions when the objective is switched from the traditional approach of flood mitigation to total benefits maximisation. This analysis allows a clear visualisation of which measures are best in each situation, showing that optimal solutions will prefer grey infrastructure when the objective is only to mitigate floods, but will prefer GBI if the objective of maximising co-benefits is added.

Further work is needed on methods for economic valuation of co-benefits such as liveability and aesthetics enhancement, biodiversity improvements and recreation. This is important considering that economic calculations are nowadays insufficient to fully represent the co-benefits related to green infrastructure in cities, since many important co-benefits are difficult to assess economically (Elmqvist et al., 2015). An improvement on economic representation of these benefits will help to encourage further application of GBI in urban spaces. Besides, this work, and most of the publications examined, which study the multiple benefits provided by GBI, focus on its positive aspects. However, these measures can also have negative impacts, also called dis-benefits or co-costs (Demuzere et al., 2014; Calliari et al., 2019), which should be quantified and considered in the analysis when assessing and comparing different alternatives. This will allow more realistic results and avoid future negative impacts, which can damage even more the acceptance of this approach. Lastly, the results obtained in this work were not discussed with the involved stakeholders. This is an important step to be performed in the future in order to validate the model outputs. Validation is particularly important for the multi-criteria analysis results, since this step determines which measures are selected to be further analysed. The not corroboration of this result can lead to the selection of measures which, for instance, have not local acceptance or which are not applicable due to particular circumstances not considered in the analysis.

5.5 Conclusions

A method to assess the performance of different green, blue and grey measures and their combinations in the achievement of flood risk reduction and the improvement of other benefits has been described and applied in this study. To achieve this, a hydrodynamic model was coupled with an evolutionary optimisation algorithm to evaluate and optimise preselected green-blue-grey measures. We also analysed how the effectiveness of optimal solutions regarding the primary function of flood risk reduction varies when the objectives are changed. This was performed applying the optimisation framework twice. First it was applied with the objectives of cost minimisation and flood risk reduction maximisation. Secondly the objectives were costs minimisation and total benefits maximisation. This allowed us to evaluate in a quantitative way the trade-offs when different benefits are pursued in stormwater infrastructure planning. Finally, we analysed how the composition of optimal solutions changes when the pursued objective is switched. In other words, how green, blue and grey measures are selected in different cases. It allows to understand which measures are best for each objective.

The results obtained can be summarised as:

- We confirmed optimisation as a helpful decision-making tool for stormwater management when several strategies are considered. More specifically, it allows to compare among optimal combinations of green, blue and grey measures for a wide range of costs. Using this approach, the decision maker can visualise complex trade-off between cost, flood damage reduction and co-benefits. Hence, the effectiveness of solutions for different levels of investment can be assessed.

- The combination of green, blue and grey measures is the best strategy in this case. This is particularly important when several benefits are considered simultaneously in urban spaces, where there are space limitations. The combination of measures allows to maximise the efficiency, with some measures performing best at flood risk reduction (grey) and other at co-benefits provision (green-blue).

- From the analysis of results with primary benefits as objective versus total benefits as objective, we conclude that there are inevitable trade-offs among different benefits obtained from different green-blue-grey measures. Our results stress the importance of considering the co-benefits as a central objective when selecting flood mitigation options. When only flood risk reduction is considered, even if green-blue infrastructure is applied, the achievement of co-benefits will be much lower. In order to achieve this, a decision process to establish priorities among benefits, or the relative importance between flood management and co-benefits, should be further studied to establish objective weights within the framework.

Even though the quantitative results in this work are indicative and uncertainty should be further assessed, we recommend the application of this type of multifunctional and multisystem assessment to support urban sustainability planning. It allows a broad and reliable comparison of diverse green-blue-grey solutions and its multiple benefits.

6

OUTLOOK

6.1 INTRODUCTION

This chapter summarises the findings of the previous chapters and discusses them in relation to the research questions presented in Chapter 1. Then, the results obtained are analysed regarding their strengths and limitations, providing critical reflections about the outcomes of this work. Finally, this chapter provides an outlook on the topic, identifying further challenges and opportunities for improvement which should be included in future research efforts.

6.2 OUTCOMES

6.2.1 Multiple-criteria analysis for GBI screening

The challenges of urbanisation and climate change are also the opportunity to develop well-planned adaptation. This presents a chance to change the approach of flood risk management and to focus on solving several issues simultaneously, achieving multiple benefits. However, drainage systems are considered complex systems, which induce difficulties for decision makers. There are several key elements that are considered either scarcely or not at all in previous works about the selection of measures, such as the impact of combining GBI and grey measures, the criteria to define the best combination of these measures in each case, and the inclusion of multiple benefits among these criteria.

In Chapter 2, a multi-criteria approach for selecting flood management measures considering the achievement of multiple benefits has been presented. It is intended to support decision makers in coping with this complex problem, having adequate flood risk management as its primary objective, but also integrating the analysis of several issues to better use the available space when implementing urban infrastructure. This multi-criteria approach helps to select green-blue and grey measures to manage different types of flood risks. This is seen as an important contribution since several studies have already suggested the importance of considering a combination of GBI with the traditional grey approach (Kabisch *et al.*, 2017a, Xie *et al.*, 2017, Dong *et al.*, 2017; Browder *et al.*, 2019).

Moreover, this methodology takes into consideration local physical characteristics and needs of the site, which allow locally-relevant solutions to be developed. This has been stated by previous works as an important factor to improve GBI acceptability by stakeholders (Davis *et al.*, 2015; Yazdanfar and Sharma, 2015). Besides, it is easy to apply since the user only has to answer simple questions. The method uses these answers to build a ranking of measures which will help the selection among them. This facilitates its use by a diverse range of users with different backgrounds which is also seen as a key factor to avoid fragmentation and to enhance GBI implementation (Matthews *et al.*, 2015; Brink *et al.*, 2016; Hoang and Fenner, 2016). Finally, this approach provides a qualitative/quantitative assessment based on indicators to assess flood management infrastructures from three perspectives: flood risk reduction, cost reduction and the provision of co-benefits. This is an important feature of the approach since many co-benefits are not easily represented in monetary terms (EEA, 2016; Kabisch *et al.*, 2017b).

The results obtained in Chapter 2 answer our first research question: How can multiple criteria, including the achievement of co-benefits, be integrated into a framework for selection of measures? Several criteria are integrated into this framework, comprising the applicability of measures for different flood types, feasibility of the application of measures according to local physical conditions, and the enhancement of different co-benefits. Also, criteria regarding cost and flood management are included. The approach enables a short and ordered list of suitable measures to be obtained.

6.2.2 Preferred co-benefits

The perspectives of different policy sectors and a wide range of stakeholders need to be considered when making decisions regarding urban infrastructure development. These elements are central to increasing GBI acceptance and to better understanding the drivers and barriers for its application. Besides, the development of multi-functional spaces and human well-being are closely related to the concept of ecosystem services. Hence, an effective way to improve the dialogue among stakeholders is obtained through the approach of ecosystem services (Everard, 2012).

In the first part of Chapter 3, an analysis linking urban ecosystem services, GBI and co-benefits is provided. The objective of this analysis is to clarify how ecosystem services can be provided by applying GBI and how this impacts on people's well-being through the provision of benefits. Benefits are defined as positive impacts, including the capacity to avoid flood damage when an option is implemented, which is our primary benefit.

The main objective in the second part of Chapter 3 was to understand how preferred benefits differ among different groups of stakeholders and how this impacts the selection of measures. This is seen as an important step to show the importance of identifying preferred benefits and the importance of multiple stakeholder participation. In this work, the benefits in addition to flood protection (or co-benefits) are divided into three groups: economic, social and environmental benefits. Through this grouping we simplified the communication with stakeholders.

Stakeholders were also divided into three groups: the general public, policy makers and the scientific community. The results obtained showed that each group has different interests and chose different co-benefits as the most important to be achieved through GBI implementation. These results show that the common practice of making decisions from a unilateral point of view, by the scientific community or policy makers, does not allow the preferences of residents or visitors with a much more local perspective to be considered. This analysis demostrates the importance of including different types of stakeholders when making decisions regarding urban infrastructure development.

The analysis performed in Chapter 3 allows us to answer our second research question: Which co-benefits are likely to be preferred by different groups of stakeholders and how can these preferences affect the selection of measures? The differences among co-benefits preferred by different stakeholders are clearly observed in the results obtained in Chapter 3. Moreover, the impacts of these differences on the selection of measures is also observed through these results. This stresses that it is not enough to consider co-benefits when planning urban infrastructure, an analysis of locally-relevant co-benefits has to be performed to choose the most adequate measures.

6.2.3 Evaluating co-benefits

The economic analysis of co-benefits can make a significant impact on decision-making, since it enables the cost-effectiveness of investments on GBI as urban flood risk management infrastructure to be better visualized. In Chapter 4, the economic viability of different flood management strategies with and without a consideration of co-benefits was analysed. This was performed calculating the monetary values of flood damage reduction, co-benefits, and costs for different combinations of green-blue-grey measures. Even though the consideration of all benefits is recommended (Kabisch *et al.*, 2017b), we only considered benefits which can be straightforwardly presented in an economic way. Since the objective was to compare results with and without benefits, even without considering all the benefits gained through the applied measures, was enough to visualize the differences and tendencies among measures.

The results obtained from this analysis show that the selection of the most cost-efficient measures changes when co-benefits are considered. While only grey measures appear as feasible if only flood damages reduction is considered, combinations of green-blue and grey measures are economically viable if co- benefits are included. Despite the uncertainties and constraints faced for calculating the costs and benefits, similar results have been found in other works (Ossa-Moreno *et al.*; 2017; Engström *et al.*, 2018).

The results obtained in Chapter 4 answer our third research question: How does the value of co-benefits affect the assessment of adaptation options? The impact of considering the economics of co-benefits is considerable, changing the selection of cost-effective measures from grey options to green-blue-grey alternatives. This was shown and further described through the result obtained in Chapter 4.

6.2.4 Analysing trade-offs

Urban drainage systems are complex systems and the combination of different strategies to achieve multiple benefits makes the decision-making process even more complex. The work presented in Chapter 5 focuses on a method to compare different combinations of green-blue and grey measures according to their cost-efficiency, searching for optimal strategies which maximize benefits and minimize costs. The method is applied twice with different benefits as the objective: first, maximizing the reduction of flood damage, and secondly, maximizing the achievement of total benefits, which is flood damage reduction plus enhancing the co-benefits.

Although the number of alternatives to assess was reduced using the method presented in Chapter 2, and despite using a 1D-1D hydrodynamic model, the application of optimisation was still a highly time-consuming process. Nevertheless, the results obtained confirmed the usefulness of combining hydrodynamic models with optimisation techniques as a decision-making tool for stormwater management. It was particularly useful in this case to enable the differences between strategies using various green- blue and grey measures to be visualised, when different objectives are pursued. Using this approach, decision makers can recognise the most effective solutions covering a wide range of costs. Moreover, they can distinguish the effectiveness achieved for each level of investment.

We have shown that a combination of green-blue and grey measures is the best option for the case studied when several benefits are pursued. In urban environments, where space is limited, the combination of green-blue and grey measures allows the efficiency to be maximised, with some measures performing best at flood damage reduction (grey) and other at co-benefit enhancement (green-blue). Our results also highlight the importance of considering the achievement of co-benefits as a relevant objective when assessing flood management options. When enhancing co-benefits is not established as an objective, even if green-blue infrastructure is used, these benefits will be achieved as a side effect and to a lesser extent.

The results presented in Chapter 5 respond to our last research questions: What are the trade-offs between flood risk reduction, cost and co-benefits? Does the enhancement of co-benefit decrease the efficiency of flood risk reduction when selecting adaptation strategies? The trade-offs between benefits and cost are visualised in the results presented in Chapter 5. In particular, the compromises between flood risk reduction and co-benefits when applying green-blue and grey measures can be observed. This shows that attention has to be paid to the decrease of efficiency on flood risk reduction when the achievement of co-benefits is also seen as a main objective.

6.3 REFLECTIONS

6.3.1 Strengths and limitations of the multi-criteria framework

The main strength of the method presented in Chapter 2 is its capacity to rank green-blue and grey measures using a holistic approach. It is a novel method which combines local characteristics, different types of flood reduction, costs and co-benefits. This method is seen as an essential step to provide pre-screening of options before the application of more complex assessment frameworks. Through this screening the number of options to be further evaluated can be much reduced, improving the efficiency of the assessment while ensuring that the final solution will be better accepted for implementation.

However, several limitations are identified regarding the development and application of this method. First, the assessment of measures regarding their performance to achieve different benefits is based on qualitative data obtained from a literature review. Projects implementing GBI are taking place in many places around the globe, so it is expected that more quantitative data is being produced. The use of data obtained from practical examples could improve the reliability of this approach regarding the performance of measures to provide different benefits. This additional data could be easily introduced into the multi-criteria analysis, being the methodology still usefull and providing perhaps more realistic results.

Second, the method was applied in three study areas showing promising results, however a validation process of the results obtained is missing. This validation should be performed in a second round of meetings with decision makers, to gain feedback about the adequacy of the measures selected. Through this process we will know if the results obtained using the methodology are suitable and if planners find the approach useful.

Finally, this method provides an ordered short-list of applicable measures. There is a missing step between this list and the definition of which options to analyse further. In this step, it should be established which green-blue and grey measures, and which combinations of them, should be selected to be further studied using more complex methods, such as hydrodynamic models and cost-benefit analysis. To achieve this, technical decision makers and local stakeholders should work together, discussing which measures will be more readily accepted for implementation, and which will have less institutional and public questioning regarding maintenance, ownership and negative impacts. Again, this implies further meetings and discussion among the actors involved for each particular case.

6.3.2 Improving the integration of multiple views

Even though the results presented in Chapter 3 help to understand how the preferences about co-benefits differ among different stakeholders, this understanding could be improved through a wider survey. The case here studied is a tourist place (Ayutthaya, Thailand) which implies some peculiarities not applicable in other cases, such as the importance of including tourists in the study. This analysis should be repeated in different study areas, predominantly residential or commercial, for instance. It would enable the results obtained in this work to be confirmed, as well as a search for different patterns of co-benefits preferences if that is the case.

More examples and a deeper analysis of results about preferences of the scientific community and technical decision makers could help to better understand the fragmentation among different disciplines. If results are grouped in different disciplines to explore how the selection of relevant co-benefits differs among them, it would allow constraints and opportunities for interdisciplinary work to be discovered. For instance, Davis *et al.* (2015) mentions that it is needed to bring together the worlds of water managers and spatial planners. A better understanding of the preferences of each group when planning urban infrastructure, will help to find common ground for interdisciplinary work. Only then would it be possible to plan cost-effective solutions to resolve several challenges at the same time.

Lastly, more extended analysis of the effects of considering different co-benefits on the selection of measures is needed. This will allow us to better understand the impacts of limited stakeholder involvement on long-term planning. In particular, it will help us to understand the consequences of not considering local residents and how this affects the acceptance of the strategy selected for implementation. Currently, more effort is being put on the analysis of stakeholders' perception regarding co-benefits to identify win-win strategies and solve GBI implementation problems (Pagano *et al.*, 2019). Hence, this is an on going line of research and a better understanding of this topic is expected in the short term.

6.3.3 Widening the type of co-benefits considered

Since the analysis presented in Chapter 4 is based on a literature review and local and regional data for the case studied here, this study does not attempt to provide precise cost and benefit data. The objective was to show how applying a holistic approach can change the selection of solutions for urban flood mitigation. In this work, only case-relevant and easily quantifiable co-benefits were considered, resulting in relatively low values of co-benefits. However, decision-making processes should also take into consideration non-monetary criteria, such as health enhancement, human well-being, liveability improvements and conservation of natural resources (EEA, 2016; Kabisch *et al.*, 2017b). More work on the quantification of these benefits is needed, developing general values according for instance to local circumstances. Thus, decision makers can use these values and do not have to conduct surveys such as willingness to pay every time they consider co-benefit quantification. Better data to quantify a wider range of co-benefits would improve the results and validity of the approach presented in this work.

This work, and most of the publications examined, which study the multiple benefits provided by GBI, focus on its positive aspects. However, these measures can also have negative impacts, which should be quantified and considered in the analysis for strategies selection. Demuzere et al. (2014) mention several of these possible negative aspects from green infrastructure, for example water pollution due to the use of fertilizers to maintain greenery, and more shade from a larger number of trees that can increase energy consumption for heating in cold weather. Also, the increased number of trees can reduce wind, leading to more air pollution at street level. Moreover, the increase in green spaces can impact on more insects and animals which can imply a risk to people mainly in tropical weather areas, which can lead to the use of pesticides polluting air and water. Finally, the management of water at the surface can involve health risk for people since surfaces in contact with stormwater can stay polluted after a storm, for instance in cases of illegal sewerage connection to the stormwater drainage system. The scarce reference of these negative elements in the literature about this topic is understandable since most of it aims to encourage the application of GBI. However, these negative aspects should also be considered when assessing and comparing different alternatives. This will allow more realistic results and avoid future negative impacts, wich can damage even more the acceptance of this approach.

6.3.4 Achieving a better visualisation of trade-offs

Although GBI are good at delivering multiple benefits simultaneously (European Commission, 2012a), these measures are not as good at managing extreme rainfall events (Zölch *et al.*, 2017). Therefore, trade-offs between the primary benefit of flood management and co-benefits are expected when selecting among GBI and grey measures. However, not much attention had been paid to this in previous works. The importance of considering these trade-offs is stressed in the results presented in Chapter 5, showing that even if a strategy achieves the highest total benefit, attention has to be paid to the resulting compromise in flood damage reduction. This outcome suggests the need to determine the importance for local stakeholders of each benefit as well as their accepted level of flood risk.

In this work the hydrodynamic model was calibrated against a well-accepted 1D-2D model that has been used for a long time. The reason behind this was lack of field data. However, the analysis would be much improved if the model was calibrated using real data. Besides, a validation process is missing in this case. When possible, hydrodynamic models should be calibrated and validated using field data, which will reduce the uncertainty about the model's results and gain the confidence of decision makers. Furthermore, it was not possible to calibrate the model after the inclusion of green infrastructure. The addition of green infrastructure in hydrodynamic models is relatively recent. In this work, parameters recommended in the literature were used. However, since the implementation of these measures has been growing in the last few years, more work should be carried out to verify these parameters using practical examples.

Finally, although the method used in Chapter 5 is good for comparison between strategies and helps the visualization of trade-offs among cost and benefits, a simpler presentation of results would improve communication with stakeholders. Usually, presenting results through Pareto fronts or statistical analysis is not the best way to successfully communicate results to decision makers. The use of maps to show optimal distributions of measures and the use of simpler plots could facilitate the discussion between the scientific community and practical decision makers. This is something that is usually neglected in scientific studies but it is also something that could do much to reduce the gap between science and practice. Furthermore, it can help to increase the acceptance of multi-purpose GBI for implementation.

6.4 THE WAY FORWARD

A change of approach on how we plan the development of flood risk management infrastructure is needed to develop safe urban spaces, capable of providing human well-being nowadays and in the future. Problems such as flood risk, heat stress, air pollution and water scarcity are likely to occur even more in the future. Since space in cities and economic resources are limited, decision makers face an enormous challenge. The use of multi-purpose infrastructures can help to address this issue, and GBI fulfils this requirement providing many benefits at the same time. However, the challenge is even larger considering that we need to act now since it will take years to implement this type of infrastructure. Hence, it will be too late if we wait until the effects of climate change are already fully detectable. This work contributes to the improvement of the planning process for flood risk management infrastructure. However, several issues still need to be addressed to continue narrowing the gap between science and practice regarding GBI implementation.

More applied research is needed in this topic. Through the use of data collection in pilot studies for example, the efficiency of these measures could be demonstrated. This could help effective communication with decision makers since it would show results in a practical way. Moreover, not only uncertainties about performance could be clarified, but also uncertainties about costs and maintenance methods. This is important since a big concern of urban drainage and sewerage authorities is the change in the approach to operation and maintenance (Davis et al., 2015). GBI implies a change from traditional approaches exclusively maintaining pipes and pumps to adding green space maintenance.

Furthermore, extra attention and much better communication should focus on the promotion of successful cases of GBI implementation in urban spaces. There are several of these cases around the world (Liu and Jensen, 2018; Hölscher et al., 2019; Miller and Montalto, 2019). These 'real world' examples could be better trusted by decision makers, complementing scientific results based on models and theory. Dissemination about positive outputs and lessons learned could greatly improve GBI acceptance for implementation by decision makers in other cities. However, most of the examples are from the developed world. It is also necessary to study how to translate these results to decision makers in developing countries, where different challenges such as lack of data can arise (De Risi et al., 2018). This is important since it is in the developing world, and especially in coastal regions, that the largest urban population growth is expected in the future (United Nations, 2014).

The importance of considering flexibility when performing economic analysis for flood risk infrastructure planning has already been discussed in previous works (Gersonius et *al.*, 2012; 2013; Babovic *et al.*, 2018; Babovic and Mijic, 2019). Further research is needed to understand the impact of including flexibility capacity in economic analysis when multiple benefits are considered. This can demonstrate to a large extent the low-regret advantage of GBI, since due to the capacity of delivering multiple benefits these measures can be cost-effective under different future scenarios. For instance, if higher rainfall intensities are not occurring then the benefit of flood risk reduction will be lower, but higher benefits can be obtained from heat stress or water scarcity management. This is an important consideration when making a long-term plan. Decision makers have no choise but to cope with high uncertainty regarding future climate conditions. Thus, demonstrating that GBI can deliver cost-effective results under different conditions can help to improve its acceptance.

Finally, more research is needed to study 'from whom' and 'to whom' are the costs and benefits of GBI. This is significant information for decision makers when thinking about the practical implementation of GBI. Several of these infrastructures can be applied in private places, such as rain barrels, green roofs, pervious pavements and rain gardens. In these cases some costs may be indirect costs, assumed through subsidies and incentives needed to encourage the application of these measures at private level. Similarly, part of the benefits from energy or water savings are expressed as bill reduction, which will be a benefit for the consumer. However, improvements in water quality or groundwater recharge are benefits which will be experienced generally rather than by each consumer. The analysis of these elements is scarce among scientific studies (Ossa-Moreno *et al.*, 2017), hence more research is needed to assess the economic impacts of GBI considering these aspects.

This is a fascinating topic with a great quantity of work carried out in the last few years, both in scientific studies and practical applications. However, a lot of work is still needed to improve knowledge, to enhance decision-making processes, to encourage measure implementation, and to evaluate its (positive and negative) impacts. The effect of applying these measures can be huge, making the difference in the attempt to achieve sustainable and safe urban spaces. A better dissemination of successful and unsuccessful experiences of GBI applications in different cities around the globe is crucial. The effects and issues from the broad application of these measures take time to fully develop. Effective ways of communicating lessons learned are needed in order to be ready for the expected as well as unknown future challenges.

The research presented in this thesis addressed several questions, but by far did not answer all the questions on how to help the consideration of GBI in flood mitigation decision-making. It developed and tested novel approaches, but also made more awareness of gaps and weaknesses to be addressed in follow up studies. The importance of practical data and validation have also been highlighted in this research. It confirms that scientific curiosity drives the fundamental research, and the search of knowledge on practical problems are essential for bringing solutions through applied research.

REFERENCES

Akbari H, Pomerantz M, Taha H. 2001. Cool surfaces and shade trees to reduce energy use and improve air quality in urban areas. *Solar Energy* **70** (3): 295–310 DOI: 10.1016/S0038-092X(00)00089-X

Al-rubaei AM, Stenglein AL, Viklander M, Blecken G. 2013. Long-Term Hydraulic Performance of Porous Asphalt Pavements in Northern Sweden. *J. Irrig. Drain Eng. (ASCE)* **139** (6): 499–505 DOI: 10.1061/(ASCE)IR.1943-4774.0000569.

Albert C, Schröter B, Haase D, Brillinger M, Henze J, Herrmann S, Gottwald S, Guerrero P, Nicolas C, Matzdorf B. 2019. Addressing societal challenges through nature-based solutions: How can landscape planning and governance research contribute? *Landscape and Urban Planning* DOI: 10.1016/j.landurbplan.2018.10.003

Alves A, Gersonius B, Kapelan Z, Vojinovic Z, Sanchez A. 2019. Assessing the Co-Benefits of green-blue-grey infrastructure for sustainable urban flood risk management. *Journal of Environmental Management* **239** (December 2018): 244–254 DOI: 10.1016/j.jenvman.2019.03.036

Alves A, Gersonius B, Sanchez A, Vojinovic Z, Kapelan Z. 2018a. Multi-criteria Approach for Selection of Green and Grey Infrastructure to Reduce Flood Risk and Increase CO-benefits. *Water Resources Management* **32** (7): 2505–2522 DOI: 10.1007/s11269-018-1943-3

Alves A, Gómez JP, Vojinovic Z, Sánchez A, Weesakul S. 2018b. Combining Co-Benefits and Stakeholders Perceptions into Green Infrastructure Selection for Flood Risk Reduction. *Environments* **5** (2): 29 DOI: 10.3390/environments5020029

Alves A, Sanchez A, Gersonius B, Vojinovic Z. 2016a. A Model-based Framework for Selection and Development of Multi-functional and Adaptive Strategies to Cope with Urban Floods. In *Procedia Engineering*. DOI: 10.1016/j.proeng.2016.07.463

Alves A, Sanchez A, Vojinovic Z, Seyoum S, Babel M, Brdjanovic D. 2016b. Evolutionary and Holistic Assessment of Green-Grey Infrastructure for CSO Reduction. *Water* **8** (9): 402 DOI: 10.3390/w8090402

Arnbjerg-Nielsen K, Willems P, Olsson J, Beecham S, Pathirana A, Bülow Gregersen I, Madsen H, Nguyen VT V. 2013. Impacts of climate change on rainfall extremes and urban drainage systems: A review. *Water Science and Technology* **68** (1): 16–28 DOI: 10.2166/wst.2013.251

Babovic F, Mijic A. 2019. Economic evaluation of adaptation pathways for an urban drainage system experiencing deep uncertainty. *Water (Switzerland)* **11** (3) DOI: 10.3390/w11030531

Babovic F, Mijic A, Madani K. 2018. Decision making under deep uncertainty for adapting urban drainage systems to change. *Urban Water Journal* **15** (6): 552–560 DOI: 10.1080/1573062X.2018.1529803

Bana e Costa CA, Antano da Silva P, Correia FN. 2004. Multicriteria Evaluation of Flood Control Measures: The Case of Ribeira do Livramento. *Water Resources and*

Management **18**: 263–283 DOI: 10.1023/B:WARM.0000043163.19531.6a

Barreto W, Vojinovic Z, Price R, Solomatine D. 2010. Multiobjective Evolutionary Approach to Rehabilitation of Urban Drainage Systems. *Journal of Water Resources Planning and Management*: 547–554 DOI: 10.1061/(ASCE)WR.1943-5452.0000070

Behroozi A, Niksokhan MH, Nazariha M. 2018. Developing a simulation-optimisation model for quantitative and qualitative control of urban run-off using best management practices Case study. **11**: 340–351 DOI: 10.1111/jfr3.12210

Berghage R, Beattie D, Jarrett AR, Thuring C, Razaei F, O'Connor TP. 2009. Green Roofs for Stormwater Runoff Control Green Roofs for Stormwater Runoff Control

Bianchini F, Hewage K. 2012. Probabilistic social cost-benefit analysis for green roofs: A lifecycle approach. *Building and Environment* **58**: 152–162 DOI: 10.1016/j.buildenv.2012.07.005

Bissonnette JF, Dupras J, Messier C, Lechowicz M, Dagenais D, Paquette A, Jaeger JAG, Gonzalez A. 2018. Moving forward in implementing green infrastructures: Stakeholder perceptions of opportunities and obstacles in a major North American metropolitan area. *Cities* **81** (August 2017): 61–70 DOI: 10.1016/j.cities.2018.03.014

Brink E, Aalders T, Ádám D, Feller R, Henselek Y, Hoffmann A, Ibe K, Matthey-Doret A, Meyer M, Negrut NL, et al. 2016. Cascades of green: A review of ecosystem-based adaptation in urban areas. *Global Environmental Change* **36**: 111–123 DOI: 10.1016/j.gloenvcha.2015.11.003

Browder G, Ozment S, Rehberger Bescos I, Gartner T, Lange G-M. 2019. *INTEGRATING GREEN AND GAY: Creating Next Generation Infrastructure.* World Bank and World Resources institute: Washington, DC.

Calliari E, Staccione A, Mysiak J. 2019. An assessment framework for climate-proof nature-based solutions. *Science of the Total Environment* **656**: 691–700 DOI: 10.1016/j.scitotenv.2018.11.341

Cardona OD, van Aalst MK, Birkmann J, Fordham M, McGregor G, Perez R, Pulwarty RS, Schipper EL., Sinh BT. 2012. Determinants of Risk: Exposure and Vulnerability. Cambridge, UK, and New York, NY, USA. DOI: 10.2307/1221563

Carter T, Keeler A. 2008a. Life-cycle cost – benefit analysis of extensive vegetated roof systems. *Journal of Environmental Management* **87**: 350–363 DOI: 10.1016/j.jenvman.2007.01.024

Carter T, Keeler A. 2008b. Life-cycle cost-benefit analysis of extensive vegetated roof systems. *Journal of Environmental Management* **87** (3): 350–363 DOI: 10.1016/j.jenvman.2007.01.024

Casal-Campos A, Fu G, Butler D, Moore A. 2015. An Integrated Environmental Assessment of Green and Gray Infrastructure Strategies for Robust Decision Making. *Environmental Science and Technology* **49** (14): 8307–8314 DOI: 10.1021/es506144f

Center for Neighborhood Technology. 2010. The Value of Green Infrastructure: A Guide to Recognizing Its Economic, Environmental and Social Benefits Available at:

http://www.cnt.org/sites/default/files/publications/CNT_Value-of-Green-Infrastructure.pdf [Accessed 13 February 2016]

Central Bureau of Statistics. 2009. Statistical Yearbook of the Netherlands Antilles 2009

Centrale Bank Curaçao en Sint Maarten. 2017. The Economy of Curaçao and Sint Maarten in Data and Charts

Charlesworth SM. 2010. A review of the adaptation and mitigation of global climate change using sustainable drainage in cities. *Journal of Water and Climate Change* **1** (3): 165–180 DOI: 10.2166/wcc.2010.035

Cheng C, Ryan RL, Warren PS, Nicolson C. 2017. Exploring Stakeholders' Perceptions of Urban Growth Scenarios for Metropolitan Boston (USA): The Relationship Between Urban Trees and Perceived Density. *Cities and the Environment* **10** (1)

Cheng M, Zhen JX, Shoemaker L. 2009. BMP decision support system for evaluating stormwater management alternatives. *Frontiers of Environmental Science & Engineering in China* **3** (4): 453–463 DOI: 10.1007/s11783-009-0153-x

Chenoweth J, Anderson AR, Kumar P, Hunt WF, Chimbwandira SJ, Moore TLC. 2018. The interrelationship of green infrastructure and natural capital. *Land Use Policy* **75**: 137–144 DOI: 10.1016/j.landusepol.2018.03.021

Chocat B, Ashley R, Marsalek J, Matos MR, Rauch W, Schilling W, Urbonas B. 2007. Toward the Sustainable Management of Urban Storm-Water. *Indoor and Built Environment* **16** (3): 273–285 DOI: 10.1177/1420326X07078854

Chow J, Savi D, Fortune D, Kapelan Z, Mebrate N. 2014. Using a systematic , multi-criteria decision support framework to evaluate sustainable drainage designs. **70** (0): 343–352 DOI: 10.1016/j.proeng.2014.02.039

Chow J, Savic D, Fortune D, Kapelan Z, Mebrate N. 2013. Translating Legislative Requirements and Best Practice Guidance into a Systematic , Multi-Criteria Decision Support Framework for Effective Sustainable Drainage Design Evaluation Conveyance & Storage New (Green Infrastructure): Source Control & Treatmen. *Proceedings of 2013 IAHR World Congress*

CIRIA. 2013. Demonstrating the multiple benefits of SuDS – A business case (Phase 2) Available at: http://www.susdrain.org/resources/ciria-guidance.html

Claus K, Rousseau S. 2012. Public versus private incentives to invest in green roofs: A cost benefit analysis for Flanders. *Urban Forestry and Urban Greening* **11** (4): 417–425 DOI: 10.1016/j.ufug.2012.07.003

Commission of the European Communities. 2009. Adapting to climate change: Towards a European framework for action

CRC for Water Sensitive Cities. 2016. Enhancing the economic evaluation of WSUD. *CRCWSC Research Synthesis* Available at: https://watersensitivecities.org.au/wp-content/uploads/2016/12/IdeasforSA_EnhancingtheEconomic_WEB.pdf

Dagenais D, Thomas I, Paquette S. 2017. Siting green stormwater infrastructure in a neighbourhood to maximise secondary benefits: lessons learned from a pilot project. *Landscape Research* **42** (2): 195–210 DOI: 10.1080/01426397.2016.1228861

Davis M, Krüger I, Hinzmann M. 2015. Coastal Protection and Suds-Nature-Based

Solutions Available at: www.recreate-net.eu

Deb K, Pratap A, Agarwal S, Meyarivan T. 2002. A Fast and Elitist Multiobjective Genetic Algorithm: NSGA-II. *IEEE Transactions on evolutionary computation* **6** (2): 182–197 DOI: 10.1109/4235.996017

Defra. 2007. An introductory guide to valuing ecosystem services. London, UK. Available at: www.defra.gov.uk

DEFRA. 2016. Local Action. *Local Action Toolkit* Available at: http://urbanwater-eco.services/toolbox/

Delelegn SWW, Pathirana A, Gersonius B, Adeogun AGG, Vairavamoorthy K. 2011. Multi-Objective Optimization of Cost-Benefit of Urban Flood Management using a 1D2D Coupled Model. *Water Science and Technology* **63** (5): 1054 DOI: 10.2166/wst.2011.290

Demuzere M, Orru K, Heidrich O, Olazabal E, Geneletti D, Orru H, Bhave AG, Mittal N, Feliu E, Faehnle M. 2014. Mitigating and adapting to climate change: Multi-functional and multi-scale assessment of green urban infrastructure. *Journal of Environmental Management* **146**: 107–115 DOI: 10.1016/j.jenvman.2014.07.025

Department for Communities and Local Government. 2009. Multi-criteria analysis : a manual. London, UK. DOI: 10.1002/mcda.399

Department of Statistics. 2011. Department of Statistics - Sint Maarten. *Census 2011* Available at: http://stat.gov.sx/ [Accessed 12 December 2017]

Derkzen ML, van Teeffelen AJA, Verburg PH. 2017. Green infrastructure for urban climate adaptation: How do residents' views on climate impacts and green infrastructure shape adaptation preferences? *Landscape and Urban Planning* **157** DOI: 10.1016/j.landurbplan.2016.05.027

Derkzen ML, Teeffelen AJA Van, Verburg PH. 2015. Quantifying urban ecosystem services based on high- resolution data of urban green space : an assessment for Rotterdam , the Netherlands: 1020–1032 DOI: 10.1111/1365-2664.12469

Dhakal KP, Chevalier LR. 2017. Managing urban stormwater for urban sustainability: Barriers and policy solutions for green infrastructure application. *Journal of Environmental Management* **203**: 171–181 DOI: 10.1016/j.jenvman.2017.07.065

Dong X, Guo H, Zeng S. 2017. Enhancing future resilience in urban drainage system: Green versus grey infrastructure. *Water Research* **124**: 280–289 DOI: 10.1016/j.watres.2017.07.038

Dorst H, van der Jagt S, Raven R, Runhaar H. 2019. Urban greening through Nature-Based Solutions – key characteristics of an emerging concept. *Sustainable Cities and Society* DOI: 10.1016/j.scs.2019.101620

EEA. 2012. Urban adaptation to climate change in Europe. Copenhagen, Denmark.

EEA. 2016. *Urban adaptation to climate change in Europe 2016. Transforming cities in a changing climate.* DOI: 10.2800/41895

Elimelech M, Phillip WA. 2011. The Future of Seawater Desalination: Energy, Technology, and the Environment. *SCIENCE* **333** (August): 712–718

Elliott a, Trowsdale S. 2007. A review of models for low impact urban stormwater drainage. *Environmental Modelling & Software* **22** (3): 394–405 DOI: 10.1016/j.envsoft.2005.12.005

Elmqvist T, Andersson E, Frantzeskaki N, McPhearson T, Olsson P, Gaffney O, Takeuchi K, Folke C. 2019. Sustainability and resilience for transformation in the urban century. *Nature Sustainability* **2**: 267–273

Elmqvist T, Setälä H, Handel SN, van der Ploeg S, Aronson J, Blignaut JN, Gómez-Baggethun E, Nowak DJ, Kronenberg J, de Groot R. 2015. Benefits of restoring ecosystem services in urban areas. *Current Opinion in Environmental Sustainability* **14**: 101–108 DOI: 10.1016/j.cosust.2015.05.001

Engström R, Howells M, Mörtberg U, Destouni G. 2018. Multi-functionality of nature-based and other urban sustainability solutions: New York City study. *Land Degradation and Development* (June): 3653–3662 DOI: 10.1002/ldr.3113

European Commission. 2012a. The Multifunctionality of Green Infrastructure. Bristol, England.

European Commission. 2012b. INTERNATIONAL COOPERATION AND DEVELOPMENTBuilding partnerships for change in developing countries. *Single Programming of Curacao, Sint Maarten,Bonaire, Sint Eustatius and Saba under 10th EDF*: 66

Everard M. 2012. Why does 'good ecological status' matter? *Water and Environment Journal* **26** (2): 165–174 DOI: 10.1111/j.1747-6593.2011.00273.x

Fenner R. 2017. Spatial evaluation of multiple benefits to encourage multi-functional design of sustainable drainage in Blue-Green cities. *Water (Switzerland)* **9** (12) DOI: 10.3390/w9120953

Fletcher TD, Shuster W, Hunt WF, Ashley R, Butler D, Arthur S, Trowsdale S, Barraud S, Semadeni-Davies A, Bertrand-Krajewski J-L, et al. 2014. SUDS, LID, BMPs, WSUD and more – The evolution and application of terminology surrounding urban drainage. *Urban Water Journal*: 1–18 DOI: 10.1080/1573062X.2014.916314

Foster J, Lowe A, Winkelman S. 2011. The Value of Green Infrastructure for Urban Climate Adaptation. *The Centre For Clean Air Policy* (February): 52

Frantzeskaki N, Mcphearson T, Collier MJ, Kendal D, Bulkeley H, Dumitru A, Walsh C, Noble K, Wyk EVAN, Ordóñez C, et al. 2019. Nature-Based Solutions for Urban Climate Change Adaptation: Linking Science, Policy, and Practice Communities for Evidence-Based Decision-Making. *BioScience* **69** (6): 455–466 DOI: 10.1093/biosci/biz034

Fratini CFF, Geldof GD, Kluck J, Mikkelsen PS. 2012. Three Points Approach (3PA) for urban flood risk management: A tool to support climate change adaptation through transdisciplinarity and multifunctionality. *Urban Water Journal* **9** (5): 317–331 DOI: 10.1080/1573062X.2012.668913

Gersonius B, Ashley R, Pathirana A, Zevenbergen C. 2013. Climate change uncertainty: building flexibility into water and flood risk infrastructure. *Climatic Change* **116** (2): 411–423 DOI: 10.1007/s10584-012-0494-5

Gersonius B, Morselt T, Nieuwenhuijzen L Van, Ashley R, Zevenbergen C, van

Nieuwenhuijzen L, Ashley R, Zevenbergen C. 2012. How the Failure to Account for Flexibility in the Economic Analysis of Flood Risk and Coastal Management Strategies Can Result in Maladaptive Decisions. *Journal of Waterway, Port, Coastal, and Ocean Engineering* **138** (5): 386–393 DOI: 10.1061/(ASCE)WW.1943-5460.0000142

Getter K, Rowe DB, Robertson GP, Cregg BM, Andresen§ JA. 2009. Carbon Sequestration Potential of Extensive Green Roofs. *Environ. Sci. Technol.* **43** (19): 7564–7570 DOI: 10.1021/es901539x

Golub D. 2014. Towards a framework for participatory flood risk assessment in urban areas with cultural heritage: The case of the Historic City of Ayutthaya, Thailand.UNESCO-IHE Institute for Water Education.

de Groot RS, Wilson M a., Boumans RMJ, Groot RS De, Wilson M a., Boumans RMJ. 2002. A typology for the classification, description and valuation of ecosystem functions, goods and services. *Ecological Economics* **41** (3): 393–408 DOI: 10.1007/s13280-014-0503-1

Haghighatafshar S, Nordlöf B, Roldin M, Gustafsson LG, la Cour Jansen J, Jönsson K. 2018. Efficiency of blue-green stormwater retrofits for flood mitigation – Conclusions drawn from a case study in Malmö, Sweden. *Journal of Environmental Management* **207**: 60–69 DOI: 10.1016/j.jenvman.2017.11.018

Haines-Young R, Potschin M. 2012. Common International Classification of Ecosystem Goods and Services (CICES): Consultation on Version 4 DOI: 10.1038/nature10650

Haines-Young RH, Potschin MB. 2010. The links between biodiversity, ecosystem services and human well-being. In *Ecosystems Ecology: A New Synthesis*Cambridge University Press; 31. DOI: 10.1017/CBO9780511750458

Hajkowicz S, Collins K. 2007. A review of multiple criteria analysis for water resource planning and management. *Water Resources Management* **21** (9): 1553–1566 DOI: 10.1007/s11269-006-9112-5

Hoang L, Fenner R a. A. 2016. System interactions of stormwater management using sustainable urban drainage systems and green infrastructure. *Urban Water Journal* **13** (7): 739–758 DOI: 10.1080/1573062X.2015.1036083

Hoang L, Fenner RA, Skenderian M. 2018. A conceptual approach for evaluating the multiple benefits of urban flood management practices. *Journal of Flood Risk Management* **11**: S943–S959 DOI: 10.1111/jfr3.12267

Hölscher K, Frantzeskaki N, McPhearson T, Loorbach D. 2019. Tales of transforming cities: Transformative climate governance capacities in New York City, U.S. and Rotterdam, Netherlands. *Journal of Environmental Management* **231** (May 2018): 843–857 DOI: 10.1016/j.jenvman.2018.10.043

Horton B, Digman CJ, Ashley RM, Gill E. 2016. BeST (Benefits of SuDS Tool) W045c BeST - Technical Guidance Release version 2

Huizinga J, De Moel H, Szewczyk W. 2017. Global flood depth-damage functions - Methodology and the database with guidelines DOI: 10.2760/16510

International Monetary Fund. 2016. KINGDOM OF THE NETHERLANDS —

CURACAO AND SINT MAARTEN. Washington, D.C.

IPCC. 2012. *Managing the Risks of Extreme Events and Disasters To Advance Climate Change Adaptation. A Special Report of Working Groups I and II of the Intergovernmental Panel on Climate Change* (and PMM C.B., V. Barros, T.F. Stocker, D. Qin, D.J. Dokken, K.L. Ebi, M.D. Mastrandrea, K.J. Mach, G.-K. Plattner, S.K. Allen, M. Tignor, ed.). Cambridge University Pres: Cambridge and New York. DOI: 10.1017/CBO9781139177245

IPPC. 2007. *Climate Change 2007: impacts, adaptation and vulnerability: contribution of Working Group II to the fourth assessment report of the Intergovernmental Panel.* DOI: 10.1256/004316502320517344

Jha AK, Bloch R, Lamond J. 2012. *Cities and Flooding: A Guide to Integrated Urban Flood Risk Management for the 21st Century.* The Word Bank. DOI: 10.1596/978-0-8213-8866-2

Jia H, Lu Y, Yu SL, Chen Y. 2012. Planning of LID–BMPs for urban runoff control: The case of Beijing Olympic Village. *Separation and Purification Technology* **84**: 112–119 DOI: 10.1016/j.seppur.2011.04.026

Jia H, Yao H, Tang Y, Yu SL, Field R, Tafuri AN. 2015. LID-BMPs planning for urban runoff control and the case study in China. *Journal of Environmental Management* **149**: 65–76 DOI: 10.1016/j.jenvman.2014.10.003

Jia H, Yao H, Tang Y, Yu SL, Zhen JX, Lu Y. 2013. Development of a multi-criteria index ranking system for urban runoff best management practices (BMPs) selection. *Environ Monit Assess* **185** (9): 7915–7933 DOI: 10.1007/s10661-013-3144-0

Jonkman SN, Brinkhuis-Jak M, Kok M. 2004. Cost benefit analysis and flood damage mitigation in the Netherlands. *Heron* **49** (1): 95–111

Kabisch N, Korn H, Stadler J, Bonn A. 2017a. *Nature - based Solutions to Climate Change Adaptation in Urban Areas.* Springer.

Kabisch N, Stadler J, Korn H, Bonn A. 2016. Nature-based solutions to climate change mitigation and adaptation in urban areas. *Ecology and Society* **21** (2): 39 DOI: 10.5751/ES-08373-210239

Kabisch N, Stadler J, Korn H, Duffield S, Bonn A. 2017b. *Proceedings of the European conference on Nature-based solutions to climate change in urban areas and their rural surroundings.* DOI: 10.19217/skr456

Karavokiros G, Lykou A, Koutiva I, Batica J, Kostaridis A, Alves A, Makropoulos C. 2016. Providing Evidence-Based, Intelligent Support for Flood Resilient Planning and Policy: The PEARL Knowledge Base. *Water* **8** (9): 392 DOI: 10.3390/w8090392

Keerakamolchai W. 2014. Towards a framework for muntifunctional flood detention facilities design in a mixed land use area The case of Ayutthaya World Heritage Site , Thailand.Asian Institute of Technology and UNESCO-IHE.

Klijn F, Mens MJP, Asselman NEM. 2015. Flood risk management for an uncertain future: economic efficiency and system robustness perspectives compared for the Meuse River (Netherlands). *Mitigation and Adaptation Strategies for Global Change* **20** (6): 1011–1026 DOI: 10.1007/s11027-015-9643-2

Kong F, Ban Y, Yin H, James P, Dronova I. 2017. Modeling stormwater management at the city district level in response to changes in land use and low impact development. *Environmental Modelling and Software* **95**: 132–142 DOI: 10.1016/j.envsoft.2017.06.021

Kosareo L, Ries R. 2007. Comparative environmental life cycle assessment of green roofs. *Building and Environment* **42** (7): 2606–2613 DOI: 10.1016/j.buildenv.2006.06.019

Kremer P, Hamstead ZA, McPhearson T. 2016. The value of urban ecosystem services in New York City: A spatially explicit multicriteria analysis of landscape scale valuation scenarios. *Environmental Science and Policy* **62**: 57–68 DOI: 10.1016/j.envsci.2016.04.012

Kuller M, Bach PM, Roberts S, Browne D, Deletic A. 2019. A planning-support tool for spatial suitability assessment of green urban stormwater infrastructure. *Science of the Total Environment* **686**: 856–868 DOI: 10.1016/j.scitotenv.2019.06.051

Lafortezza R, Chen J, van den Bosch CK, Randrup TB. 2018. Nature-based solutions for resilient landscapes and cities. *Environmental Research* DOI: 10.1016/j.envres.2017.11.038

Liquete C, Udias A, Conte G, Grizzetti B, Masi F. 2016. Integrated valuation of a nature-based solution for water pollution control. Highlighting hidden benefits. *Ecosystem Services* **22** (September): 392–401 DOI: 10.1016/j.ecoser.2016.09.011

Liu L, Jensen MB. 2018. Green infrastructure for sustainable urban water management: Practices of five forerunner cities. *Cities* **74** DOI: 10.1016/j.cities.2017.11.013

Liu Y, Ahiablame LM, Bralts VF, Engel BA. 2015. Enhancing a rainfall-runoff model to assess the impacts of BMPs and LID practices on storm runoff. *Journal of Environmental Management* **147**: 12–23 DOI: 10.1016/j.jenvman.2014.09.005

Lorphensri O., Ladawadee A., Dhammasarn S. 2011. Review of Groundwater Management and Land Subsidence in Bangkok, Thailand. In *Groundwater and Subsurface Environments*, Taniguchi M. (ed.).Springer: Tokyo.

Lovell S. T., Taylor J. R. 2013. Supplying urban ecosystem services through multifunctional green infrastructure in the United States: 1447–1463 DOI: 10.1007/s10980-013-9912-y

Lundy L, Wade R. 2011. Integrating sciences to sustain urban ecosystem services. *Progress in Physical Geography* **35** (5): 653–669 DOI: 10.1177/0309133311422464

Macmullan E, Reich S. 2007. The Economics of Low-Impact Development : A Literature Review

Maier HR, Kapelan Z, Kasprzyk J, Kollat J, Matott LS, Cunha MC, Dandy GC, Gibbs MS, Keedwell E, Marchi A, et al. 2014. Evolutionary algorithms and other metaheuristics in water resources : Current status , research challenges and future directions. *Environmental Modelling and Software* **62**: 271–299 DOI: 10.1016/j.envsoft.2014.09.013

Maier HR, Razavi S, Kapelan Z, Matott LS, Kasprzyk J, Tolson BA. 2019. Introductory overview: Optimization using evolutionary algorithms and other metaheuristics. *Environmental Modelling and Software*: 195–213 DOI: 10.1016/j.envsoft.2018.11.018

Mala-Jetmarova H, Barton A, Bagirov A. 2015. Sensitivity of algorithm parameters and objective function scaling in multi-objective optimisation of water distribution systems. *Journal of Hydroinformatics* **17** (6): 891–916 DOI: 10.2166/hydro.2015.062

Martin C, Ruperd Y, Legret M. 2007. Urban stormwater drainage management: The development of a multicriteria decision aid approach for best management practices. *European Journal of Operational Research* **181**: 338–349 DOI: 10.1016/j.ejor.2006.06.019

Di Matteo M, Maier HR, Dandy GC. 2019. Many-objective portfolio optimization approach for stormwater management project selection encouraging decision maker buy-in. *Environmental Modelling and Software* **111** (April 2017): 340–355 DOI: 10.1016/j.envsoft.2018.09.008

Matthews T, Lo AY, Byrne JA. 2015. Reconceptualizing green infrastructure for climate change adaptation: Barriers to adoption and drivers for uptake by spatial planners. *Landscape and Urban Planning* **138**: 155–163 DOI: 10.1016/j.landurbplan.2015.02.010

McPhearson T, Haase D, Kabisch N, Gren Å. 2016. Advancing understanding of the complex nature of urban systems. *Ecological Indicators* **70**: 566–573 DOI: 10.1016/j.ecolind.2016.03.054

Meerow S, Newell JP. 2016. Spatial planning for multifunctional green infrastructure: Growing resilience in Detroit. *Landscape and Urban Planning* **159**: 62–75 DOI: 10.1016/j.landurbplan.2016.10.005

Mei C, Liu J, Wang H, Yang Z, Ding X, Shao W. 2018. Integrated assessments of green infrastructure for flood mitigation to support robust decision-making for sponge city construction in an urbanized watershed. *Science of the Total Environment* **639**: 1394–1407 DOI: 10.1016/j.scitotenv.2018.05.199

Meteorological Department St. Maarten. 2018. Climatological Summary 2017: 34 Available at: http://www.meteosxm.com/wp-content/uploads/Climatological-Summary-20172.pdf [Accessed 16 July 2018]

Millennium Ecosystem Assessment. 2005. *Ecosystems and Human Well-being: Synthesis*. Island Press: Washington, DC. DOI: 10.1196/annals.1439.003

Miller SM, Montalto FA. 2019. Stakeholder perceptions of the ecosystem services provided by Green Infrastructure in New York City. *Ecosystem Services* **37** (April): 100928 DOI: 10.1016/j.ecoser.2019.100928

Montalto F, Behr C, Alfredo K, Wolf M, Arye M, Walsh M. 2007. Rapid assessment of the cost-effectiveness of low impact development for CSO control. *Landscape and Urban Planning* **82** (3): 117–131 DOI: 10.1016/j.landurbplan.2007.02.004

Moura NCB, Pellegrino PRM, Martins JRS. 2016. Best management practices as an alternative for flood and urban storm water control in a changing climate. *Journal of Flood Risk Management* **9** (3): 243–254 DOI: 10.1111/jfr3.12194

Mynett AE, Vojinovic Z. 2009. Hydroinformatics in multi-colours—part red: urban flood and disaster management. *Journal of Hydroinformatics* **11**: 166 DOI: 10.2166/hydro.2009.027

Narayanan A, Pitt R. 2006. Costs of Urban Stormwater Control Practices. Tuscaloosa. DOI: doi:10.1061/40517(2000)38

National Renewable Energy Laboratory. 2015. Energy Snapshot Saint-Martin/Sint Maarten. *Energy Transition Initiative - Islands*: 5

Naumann S, Rayment M, Nolan P, Forest TM, Gill S, Infrastructure G, Forest M. 2011. Design, implementation and cost elements of Green Infrastructure projects. Final Report

Nesshöver C, Assmuth T, Irvine KN, Rusch GM, Waylen KA, Delbaere B, Haase D, Jones-Walters L, Keune H, Kovacs E, et al. 2017. The science, policy and practice of nature-based solutions: An interdisciplinary perspective. *Science of the Total Environment* DOI: 10.1016/j.scitotenv.2016.11.106

Nicklow J, Asce F, Reed P, Asce M, Savic D, Dessalegne T, Asce M, Harrell L, Asce M, Chan-hilton A, et al. 2010. State of the Art for Genetic Algorithms and Beyond in Water Resources Planning and Management. (August): 412–432

NYC Environmental Protection. 2013. NYC Green Infrastructure Co-Benefits Calculator. *NYC Green Infrastructure Co-Benefits Calculator* Available at: http://www.nycgicobenefits.net/ [Accessed 12 August 2016]

O'Donnell EC, Lamond JE, Thorne CR. 2017. Recognising barriers to implementation of Blue-Green Infrastructure: a Newcastle case study. *Urban Water Journal* **14** (9): 964–971 DOI: 10.1080/1573062X.2017.1279190

Ossa-Moreno J, Smith KM, Mijic A. 2017. Economic analysis of wider benefits to facilitate SuDS uptake in London, UK. *Sustainable Cities and Society* **28**: 411–419 DOI: 10.1016/j.scs.2016.10.002

Pagano A, Pluchinotta I, Pengal P, Cokan B, Giordano R. 2019. Engaging stakeholders in the assessment of NBS effectiveness in flood risk reduction: A participatory System Dynamics Model for benefits and co-benefits evaluation. *Science of The Total Environment* **690**: 543–555 DOI: 10.1016/j.scitotenv.2019.07.059

Patiño Gómez J. 2017. Assessment of Green Infrastructure Measures to Reduce Stormwater Runoff and Enhance Multiple Benefits in Urban Areas.Asian Institute of Technology, Thailand; UNESCO–IHE, The Netherlands.

PEARL Project. 2015. PEARL Knowledge Base Available at: http://pearl-kb.hydro.ntua.gr/ [Accessed 1 November 2015]

Pezzaniti D, Beechman S, Kandassamy J. 2009. Influence of clogging on the effective life of permeable pavements. *Proceedings of Institution of Civil Engineering - Water Management* **162** (3): 211–220 DOI: 10.1680/wama.2009.00034

Porsche U, Kohler M. 2003. LIFE CYCLE COSTS OF GREEN ROOFS - A Comparison of Germany, USA, and Brazil - Ulrich. In *RIO 3 - World Climate & Energy Event*Rio de Janeiro; 461–467.

Qiao XJ, Kristoffersson A, Randrup TB. 2018. Challenges to implementing urban sustainable stormwater management from a governance perspective: A literature review. *Journal of Cleaner Production* **196**: 943–952 DOI: 10.1016/j.jclepro.2018.06.049

Radjouki E, Hooft Graafland C. 2014. National Energy Policy for Country Sint Maarten Towards a sustainable development

Raymond CM, Frantzeskaki N, Kabisch N, Berry P, Breil M, Nita MR, Geneletti D, Calfapietra C. 2017. A framework for assessing and implementing the co-benefits of nature-based solutions in urban areas. *Environmental Science and Policy* **77** (June): 15–24 DOI: 10.1016/j.envsci.2017.07.008

Recanatesi F, Petroselli A, Ripa MN, Leone A. 2017. Assessment of stormwater runoff management practices and BMPs under soil sealing: A study case in a peri-urban watershed of the metropolitan area of Rome (Italy). *Journal of Environmental Management* **201**: 6–18 DOI: 10.1016/j.jenvman.2017.06.024

Riabacke M, Danielson M, Ekenberg L. 2012. State-of-the-art prescriptive criteria weight elicitation. *Advances in Decision Sciences* **2012** DOI: 10.1155/2012/276584

De Risi R, De Paola F, Turpie J, Kroeger T. 2018. Life Cycle Cost and Return on Investment as complementary decision variables for urban flood risk management in developing countries. *International Journal of Disaster Risk Reduction* **28** (February): 88–106 DOI: 10.1016/j.ijdrr.2018.02.026

Roachanakanan T. 2013. Changing in drainage pattern and increasing flood risk in Thailand. In *Asia Flood Conference* Bangkok.

Rossman LA. 2010. Storm water management model. User's manual. USEPA, Cincinnati, OH. Available at: http://www.epa.gov/water-research/storm-water-management-model-swmm

Rowe Bradley D. 2011. Green roofs as a means of pollution abatement. *Environmental Pollution* **159** (8–9): 2100–2110 DOI: 10.1016/j.envpol.2010.10.029

RPA. 2004. Evaluating a multi-criteria analysis (MCA) methodology for application to flood management and coastal defence appraisals. London, UK.

Saarikoski H, Mustajoki J, Barton DN, Geneletti D, Langemeyer J, Gomez-Baggethun E, Marttunen M, Antunes P, Keune H, Santos R. 2016. Multi-Criteria Decision Analysis and Cost-Benefit Analysis: Comparing alternative frameworks for integrated valuation of ecosystem services. *Ecosystem Services* **22** (November): 238–249 DOI: 10.1016/j.ecoser.2016.10.014

Santamouris M. 2013. Using cool pavements as a mitigation strategy to fight urban heat island - A review of the actual developments. *Renewable and Sustainable Energy Reviews* **26**: 224–240 DOI: 10.1016/j.rser.2013.05.047

Santamouris M. 2014. On the energy impact of urban heat island and global warming on buildings. *Energy and Buildings* **82**: 100–113 DOI: 10.1016/j.enbuild.2014.07.022

Santoro S, Pluchinotta I, Pagano A, Pengal P, Cokan B, Giordano R. 2019. Assessing stakeholders' risk perception to promote Nature Based Solutions as flood protection strategies: The case of the Glinščica river (Slovenia). *Science of the Total Environment* **655**: 188–201 DOI: 10.1016/j.scitotenv.2018.11.116

Schifman LA, Herrmann DL, Shuster WD, Ossola A, Garmestani A, Hopton ME. 2017. Situating Green Infrastructure in Context: A Framework for Adaptive Socio-Hydrology in Cities. *Water Resources Research*: 1–16 DOI: 10.1002/2017WR020926

Shoemaker L, Riverson JJ, Alvi K, Zhen JX, Paul S, Rafi T. 2009. SUSTAIN - A Framework for Placement of Best Management Practices in Urban Watersheds to Protect Water Quality. Usepa. Available at: http://www.epa.gov/nrmrl/wswrd/wq/models/sustain/

Simonovic SP. 2009. *Managing Water Resources*. UNESCO: London, UK.

Simonovic SP. 2012. *Floods in a Changing Climate: Risk Management*. Cambridge University Press: New York.

Staub C, Ott W, Heusi F, Klingler G, Jenny A, Häcki M, Hauser A. 2011. Indicators for Ecosystem Goods and Services: Framework, methodology and recommendations for a welfare-related environmental reporting. Bern.

Stratus Consulting. 2009. A Triple Bottom Line Assessment of Traditional and Green Infrastructure Options for Controlling CSO Events in Philadelphia's Watersheds, Final Report Available at: http://www.sbnphiladelphia.org/initiatives/green_economy_task_force/resources/

Teng J, Jakeman AJ, Vaze J, Croke BFW, Dutta D, Kim S. 2017. Flood inundation modelling: A review of methods, recent advances and uncertainty analysis. *Environmental Modelling and Software* **90**: 201–216 DOI: 10.1016/j.envsoft.2017.01.006

Thorne CR, Lawson EC, Ozawa C, Hamlin SL, Smith LA. 2018. Overcoming uncertainty and barriers to adoption of Blue-Green Infrastructure for urban flood risk management. *Journal of Flood Risk Management* **11**: S960–S972 DOI: 10.1111/jfr3.12218

Tzoulas K, Korpela K, Venn S, Yli-pelkonen V, Ka A, Niemela J, James P, Ka??mierczak A, Niemela J, James P, et al. 2007. Promoting ecosystem and human health in urban areas using Green Infrastructure: A literature review. *Landscape and Urban Planning* **81** (3): 167–178 DOI: 10.1016/j.landurbplan.2007.02.001

UDFCD. 2010. *Urban Storm Drainage Criteria Manual Volume 3, Stormwater Best Management Practice* (Urban Drainage and Flood Control District, ed.). Water Resources Publications: Denver.

Udoh E, Wang FZ. 2009. *Handbook of Research on Grid Technologies and Utility Computing: Concepts for Managing Large-Scale Applications*. IGI Global. DOI: 10.4018/978-1-60566-184-1

UNDP. 2012. *Flood Risk Reduction : Innovation and technology in risk mitigation and development planning in Small Island Developing States : towards floor risk reduction in Sint Maarten*. United Nations Development Programme (UNDP).

United Nations. 2014. World Urbanization Prospects Available at: http://esa.un.org/unpd/wup/Highlights/WUP2014-Highlights.pdf [Accessed 6 October 2015]

Urrestarazu Vincent S, Radhakrishnan M, Hayde L, Pathirana A. 2017. Enhancing the Economic Value of Large Investments in Sustainable Drainage Systems (SuDS) through Inclusion of Ecosystems Services Benefits. *Water* **9**: 841 DOI: 10.3390/w9110841

USEPA. 2000. Low Impact Development (LID). A Literature Review. Washington, DC.

USEPA. 2012. Reducing Urban Heat Islands: Compendium of Strategies - Cool Pavements

Versini PA, Kotelnikova N, Poulhes A, Tchiguirinskaia I, Schertzer D, Leurent F. 2018. A distributed modelling approach to assess the use of Blue and Green Infrastructures to fulfil stormwater management requirements. *Landscape and Urban Planning* **173** (February): 60–63 DOI: 10.1016/j.landurbplan.2018.02.001

Vojinovic Z. 2015. *Floor Risk: The Holistic Perspective.* IWA Publishing.

Vojinovic Z, Sanchez A. 2008. Optimising sewer system rehabilitation strategies between flooding, overflow emissions and investment costs. In *11th International Conference on Urban Drainage*Edinburgh, Scotland, UK; 31 August–5 September.

Vojinovic Z, van Teeffelen J. 2007. An Integrated Stormwater Management Approach for Small Islands in Tropical Climates. *Urban Water Journal* **4** (3): 211 – 231 DOI: http://dx.doi.org/10.1080/15730620701464190

Vojinovic Z, Hammond M, Golub D, Hirunsalee S, Weesakul S, Meesuk V, Medina N, Sanchez A, Kumara S, Abbott M. 2016a. Holistic approach to flood risk assessment in areas with cultural heritage: a practical application in Ayutthaya, Thailand. *Natural Hazards* **81** (1): 589–616 DOI: 10.1007/s11069-015-2098-7

Vojinovic Z, Keerakamolchai W, Weesakul S, Pudar RS, Medina N, Alves A. 2016b. Combining Ecosystem Services with Cost-Benefit Analysis for Selection of Green and Grey Infrastructure for Flood Protection in a Cultural Setting. *Environments* **4** (3) DOI: 10.3390/environments4010003

Vojinovic Z, Sahlu S, Torres a. S, Seyoum SD, Anvarifar F, Matungulu H, Barreto W, Savic D, Kapelan Z. 2014. Multi-objective rehabilitation of urban drainage systems under uncertainties. *Journal of Hydroinformatics* | **16** (5): 1–18 DOI: 10.2166/hydro.2014.223

Vojinovic Z, Solomatine D, Price RK. 2006. Dynamic least-cost optimisation of wastewater system remedial works requirements: 467–475 DOI: 10.2166/wst.2006.574

Voskamp IMM, Van de Ven FHM. 2015. Planning support system for climate adaptation: Composing effective sets of blue-green measures to reduce urban vulnerability to extreme weather events. *Building and Environment* **83**: 159–167 DOI: 10.1016/j.buildenv.2014.07.018

WHO. 2016. Health and sustainable development. *Urban green spaces* Available at: ttp://www.who.int/sustainable-development/cities/health-risks/urban-greenspace/%0Aen/ [Accessed 5 December 2015]

Wihlborg M, Sörensen J, Alkan Olsson J. 2019. Assessment of barriers and drivers for implementation of blue-green solutions in Swedish municipalities. *Journal of Environmental Management* **233** (July 2018): 706–718 DOI: 10.1016/j.jenvman.2018.12.018

Wild TC, Henneberry J, Gill L. 2017. Comprehending the multiple 'values' of green infrastructure – Valuing nature-based solutions for urban water management from multiple perspectives. *Environmental Research* **158** (November 2016): 179–187 DOI: 10.1016/j.envres.2017.05.043

Woods-Ballard B, Kellagher R, Martin P, Jefferies C, Bray R, Shaffer P. 2007. The SUDS manual. London, UK, UK. DOI: London C697

Woodward M, Gouldby B, Kapelan Z, Hames D. 2014. Multiobjective Optimization for Improved Management of Flood Risk. (February): 201–215 DOI: 10.1061/(ASCE)WR.1943-5452.0000295.

World Bank. 2017. Implementing Nature Based Flood Protection DOI: 10.1596/28837

WWAP/UN-Water. 2018. *The United Nations World Water Development Report 2018: Nature-Based Solutions for Water.* UNESCO: Paris. Available at: www.unesco.org/open-access/

Xie J, Chen H, Liao Z, Gu X, Zhu D. 2017. An integrated assessment of urban flooding mitigation strategies for robust decision making. *Environmental Modelling and Software* **95**: 143–155 DOI: 10.1016/j.envsoft.2017.06.027

Yazdanfar Z, Sharma A. 2015. Urban drainage system planning and design – challenges with climate change and urbanization: a review. *Water Science & Technology* **72** (2): 165 DOI: 10.2166/wst.2015.207

Yong CF, McCarthy DT, Deletic A. 2013. Predicting physical clogging of porous and permeable pavements. *Journal of Hydrology* **481**: 48–55 DOI: 10.1016/j.jhydrol.2012.12.009

Young KD, Dymond RL, Asce M, Kibler DF, Asce M. 2011. Development of an Improved Approach for Selecting Storm-Water Best Management Practices. *Journal of Water Resources Planning and Management* **137** (June): 268–275 DOI: 10.1061/(ASCE)WR.1943-5452.0000110.

Young KD, Kibler DF, Benham BL, Loganathan G V. 2009. Application of the Analytical Hierarchical Process for Improved Selection of Storm-Water BMPs. *Journal of Water Resources Planning and Management* **135** (August): 264–275 DOI: 10.1061/(ASCE)0733-9496(2009)135:4(264)

Zhang G, Hamlett JM, Reed P, Tang Y. 2013. Multi-Objective Optimization of Low Impact Development Designs in an Urbanizing Watershed. *Open Journal of Optimization* **2** (December): 95–108 DOI: http://dx.doi.org/10.4236/ojop.2013.24013

Zhou Q, Leng G, Huang M. 2018. Impacts of future climate change on urban flood risks: benefits of climate mitigation and adaptations. *Hydrology and Earth System Sciences Discussions*: 1–31 DOI: 10.5194/hess-2016-369

Zhou Q, Leng G, Su J, Ren Y. 2019. Comparison of urbanization and climate change impacts on urban flood volumes: Importance of urban planning and drainage adaptation. *Science of the Total Environment* **658**: 24–33 DOI: 10.1016/j.scitotenv.2018.12.184

Zölch T, Henze L, Keilholz P, Pauleit S. 2017. Regulating urban surface runoff through nature-based solutions – An assessment at the micro-scale. *Environmental Research* **157** (May): 135–144 DOI: 10.1016/j.envres.2017.05.023

Zölch T, Wamsler C, Pauleit S. 2018. Integrating the ecosystem-based approach into municipal climate adaptation strategies: The case of Germany. *Journal of Cleaner Production* DOI: 10.1016/j.jclepro.2017.09.146

APPENDIX A

Measures Selection Tool Questionnaire

As part of PEARL Project, a tool to help the selection of structural measures to cope with urban floods has been developed. The aim of this tool is to help technical decision makers with good knowledge about the problem and the area. The expected result is an ordered list of applicable measures for the site considered.

The questionnaire is divided in four parts: "Selection of measures", "Suitability ranking", "Preferred benefits ranking" and "Main objective identification".

Your contribution to this survey is much appreciated. The results will be used to validate and improve the tool, allowing its use in different sites helping the development of sustainable and more adaptable flood reduction management.

*Required

INTRODUCTION

1. **How are you involved in the flood management process for the area?** *
   Tick all that apply.

   ☐ You are linked with technical decision making processes

   ☐ You are linked with the political decision making process

   ☐ You are part of the institutional sector

   ☐ You are involved on system operation

   ☐ You live in the area

   ☐ Other

2. **Do you know in detail the flood problem affecting the area?** *
   Mark only one oval.

   ◯ Yes
   ◯ In part
   ◯ No

MEASURES SELECTION

Next questions are used for screening stormwater management measures, selecting those that are applicable for the area under study.

3. **Which is/are the type/s of floods affecting the area?** *
   Tick all that apply.

   ☐ Fluvial
   ☐ Pluvial
   ☐ Costal
   ☐ Flash
   ☐ Groundwater

4. **Which is the dominant infiltration capacity of the soil?**
   (please, answer this question only in case of pluvial floods)
   Mark only one oval.

   ◯ High (less of 10 % clay and more of 90 % sand or grave, hydraulic conductivity higher than 40 micrometres per second).

   ◯ Medium (between 10 y 40 % clay and 50 a 90 % sand, hydraulic conductivity between 1 y 40 micrometres per second).

   ◯ Low (more of 40 % clay and less of 50 % sand, hydraulic conductivity lower than 1 micrometre per second)

5. **Which is the average groundwater depth in the area?**
(please, answer this question only in case of pluvial floods)
Mark only one oval.

○ Less or equal than 1m

○ Higher than 1m

○ N/A

6. **Which is the average bedrock depth in the area?**
(please, answer this question only in case of pluvial floods)
Mark only one oval.

○ Less or equal than 1m

○ Higher than 1m

○ N/A

7. **Which is the average slope of the drainage area?**
(please, answer this question only in case of pluvial floods)
Mark only one oval.

○ Less or equal than 5%

○ Higher than 5%

○ N/A

8. **Which is the type of sewer system in the area?**
(please, answer this question only in case of pluvial floods)
Mark only one oval.

○ Combined

○ Separate

○ N/A

9. **Is it possible to relocate, or rise the level, of affected assets?**
(please, answer this question only in case of pluvial floods)
Mark only one oval.

○ No

○ Yes

○ N/A

10. **Can the floods occurring in the area be considered as flash floods?**
(please, answer this question only in case of pluvial floods)
Mark only one oval.

○ Yes

○ No

○ N/A

11. **How is the urban configuration in the zone under risk?**
(please, answer this question only in case of coastal floods)
Mark only one oval.

○ Residential, highly urbanised

○ Residential, low density

○ Industrial/Commercial

○ N/A

12. **How is the space availability along the coastline?**
(please, answer this question only in case of coastal floods)
Mark only one oval.

○ Free coastline

○ Urbanised coastline

○ N/A

13. **Is the area under risk already developed?**
(please, answer this question only in case of groundwater floods)
Mark only one oval.

○ No

○ Yes

○ N/A

14. Do you think there is other relevant information regarding local characteristics?

_____

_____

_____

_____

SUITABILITY RANKING

The questions in this section are used to order the selected measures. Preference is given to measures better suited according to local conditions.

15. Which is the percentage of public spaces in the total area?
   Mark only one oval.

   ⬭ Less than 25%

   ⬭ Higher than 25%

16. Are there free linear spaces in the side of roads and sidewalks?
   Mark only one oval.

   ⬭ Yes

   ⬭ No

17. Which is the population density in the area?
   Mark only one oval.

   ⬭ Less than 100 inhabitants per hectare

   ⬭ More than 100 inhabitants per hectare

18. Which is the most important land use in the area?
   Mark only one oval.

   ⬭ Residential

   ⬭ Industrial / Commercial

19. Has the sewage system wastewater treatment plant? and/or Are combined sewer overflows (CSO) an issue in the area?
   Mark only one oval.

   ⬭ Yes

   ⬭ No

20. Do you think there is other relevant information regarding the suitability of measures according local conditions?

   _____

PREFERRED BENEFITS RANKING

Please, select scores between 1 and 10 for each benefit according to how important it is for the area (1 corresponds to low importance, and 10 corresponds to high importance).

21. WATER QUALITY IMPROVEMENT
   It is related with the general capacity of the measures to remove runoff pollutants.
   Mark only one oval.

	1	2	3	4	5	6	7	8	9	10	
Low Importance	⬭	⬭	⬭	⬭	⬭	⬭	⬭	⬭	⬭	⬭	High Importance

22. ENVIRONMENTAL BENEFITS

It is the capacity of the different measures to contribute with the space amenity, ecological diversity, groundwater recharge and water reuse.
Mark only one oval.

	1	2	3	4	5	6	7	8	9	10	
Low importance	○	○	○	○	○	○	○	○	○	○	High importance

23. LIVEABILITY ENHANCEMENT

It is the capacity of these measures to improve local aesthetics, to allow species habitat creation, and urban heat island effect reduction; while having community acceptability and low public safety risk.
Mark only one oval.

	1	2	3	4	5	6	7	8	9	10	
Low importance	○	○	○	○	○	○	○	○	○	○	High importance

24. ECONOMIC BENEFITS

It is related with the measures capacity of allowing production of food or materials, and the possibility of generate energy savings or new jobs through its application.
Mark only one oval.

	1	2	3	4	5	6	7	8	9	10	
Low importance	○	○	○	○	○	○	○	○	○	○	High importance

25. SOCIO-CULTURAL BENEFITS

It is the measures capacity of creating educational spaces, generating community engagement and spaces for recreation.
Mark only one oval.

	1	2	3	4	5	6	7	8	9	10	
Low importance	○	○	○	○	○	○	○	○	○	○	High importance

26. Please, add any comments or suggestions:

MAIN OBJECTIVE IDENTIFICATION

Please, select the option for each objective to give an idea of how important it is for the area under study.

27. **How frequent are flood problems affecting buildings and generating high damages in the area?**
Mark only one oval.

- ☐ Flood events generating important damages occur every year
- ☐ Flood events generating important damages occur around once each two years
- ☐ Flood events generating important damages occur once each 5 or more years

28. **How important is the minimisation of costs in the case under study?**
Mark only one oval.

- ☐ There is very low budget to invest on measures implementation and cost reduction is a main concern
- ☐ There are some restrictions of budget but cost reduction is not the main concern
- ☐ There is budget available for measures implementation and cost reduction is not a relevant concern

29. **How interesting is for the case the achievement of other benefits besides flood reduction?**
Mark only one oval.

- ☐ The enhancement of urban space through the achievement of co-benefits is also a main objective for this case
- ☐ The achievement of co-benefits is not a main objective for this case but measures with this target are preferred
- ☐ The enhancement of urban space through the achievement of co-benefits is not important in this case

APPENDIX B

<u>**Questionnaire to stakeholders:**</u>

1- Select the type of stakeholder that best describes your position
   - National level government
   - Local level government
   - Environmental or utility authority
   - NGO or similar
   - Researcher
   - Consultant
   - Inhabitant
   - Commercial owner
   - Tourist

2- Which is the most important environmental benefit that should be enhanced in the area by the application of green infrastructure?
   - Water quality of receiving bodies
   - Groundwater recharge
   - Biodiversity and ecology enhancement
   - Temperature reduction (heat stress reduction)
   - Air quality improvement

3- Which is the most important social benefit that should be enhanced in the area by the application of green infrastructure?
   - Amenity and aesthetics
   - Recreation and health (increment of green area per inhabitant)
   - Food security

4- Which is the most important economic benefit that should be enhanced in the area by the application of green infrastructure?
   - Rainwater harvesting
   - Pumping and treatment reduction
   - Saving energy in buildings
   - Real estate value appreciation

5- What is the level of contribution of stormwater runoff to the pollution and degradation of water bodies?
   - High
   - Medium
   - Low

6- What is the current usage of groundwater extraction and water table depletion?
   - High
   - Medium
   - Low

7- What is the need of enhancement of biodiversity and ecology in the urban area?
   - High
   - Medium
   - Low

8- What is the level of impact of heat stress effect on the population in the urban area?
- High
- Medium
- Low

9- What is the importance of improving air quality in the urban area?
- High
- Medium
- Low

10- What is the requirement of landscape improvement by including more natural spaces to enhance amenity and liveability of the community?
- High
- Medium
- Low

11- What is the importance of producing food in the area to enhance food security?
- High
- Medium
- Low

12- What is the importance of reducing water demand by using alternative water sources such as rainwater harvesting?
- High
- Medium
- Low

13- What is the importance of reducing pumping and treatment of stormwater in the area?
- High
- Medium
- Low

14- What is the importance of reducing energy consumption in buildings by reducing air conditioner and ventilation systems demand?
- High
- Medium
- Low

15- What is the importance of increasing real estate value in the urban area?
- High
- Medium
- Low

LIST OF ACRONYMS

1D	One Dimensional
2D	Two Dimensional
ACB	Annual Co-Benefits
BAU	Bussines As Usual
BMP	Best Management Practice
CBA	Cost Benefits Analysis
CIRIA	Construction Industry Research and Information Association
Co-Ben	Co-Benefits
CPI	Consumer Price Index
CRC	Cooperative Research Centre
DEFRA	Department of Environment, Food and Rural Affairs
DR	Damage Reduction
DSS	Decision Support System
EAB	Expected Annual Benefits
EbA	Ecosystem-Based Adaptation
EEA	European Environment Agency
ES	Ecosystem Services
GA	Genetic Algorithm
GBI	Green-Blue Infrastructure
GI	Green Infrastructure
GIS	Geographic Information System
GR	Green Roofs
IPCC	Intergovernmental Panel on Climate Change
KB	Knowledge Base
LID	Low Impact Development
LT	Life Time
MCDA	Multi Criteria Decision Analysis
NBS	Nature Based Solution
NGO	Non-Governmental Organization
NN	Non-dominated Solutions
NPV	Net Present Value
NSGA	Non-dominated Sorting Genetic Algorithm
NYC	New York City
O&M	Operation and Maintenance
ODB	Open Detention Basin
PB	Primary Benefit
PEARL	Preparing for Extreme And Rare events in coastaL regions
Pi	Pipes
PP	Pervious Pavements
PV	Present Value

RB	Rain Barrels
RDA	Risk and Policy Analysts
RP	Return Period
SuDS	Sustainable Drainage System
SWMM	Stormwater Management Model
TB	Total Benefits
UDFCD	Urban Drainage and Flood Control District
UNDP	United Nations Development Programme
UNESCO	United Nations Educational, Scientific and Cultural Organization
USDC	Urban Storm Drainage Criteria
USEPA	United States Environmental Protection Agency
WHO	World Health Organization
WHS	World Heritage Site
WQ	Water Quality
WSUD	Water Sensitive Urban Design
WWAP/UN-Water	United Nations World Water Assessment Programme

LIST OF TABLES

LIST OF FIGURES

ACKNOWLEDGMENTS

This is the end of a great period, full of learning and memorable experiences, for which I am most thankful. It has been four long years to accomplish this objective which would never had been possible without the people that shared with me this path. The list of people is long and surely incomplete.

I would like to thank my supervisory team, their guidance enabled me to grow as a researcher in the last four years. My promotor Damir Brdjanovic, thank you for the advice and help throughout these years. My co-promotor Zoran Vojinovic, thanks for trusting me for this position, for your guidance and creative discussions which helped me to successfully develop this topic. I would like also to thank my supervisors, Arlex Sanchez and Berry Gersonius. Arlex, thanks for being there since the very beginning, for your counsel, and for your invaluable help with clusters, computers, cables, coding, etc. which made possible to get results in a reasonable time. Berry, many thanks for the inspirational and reflective talks and suggestions, it was a great support in improving this research. Finally, I would like to thank my external supervisor Zoran Kapelan for all the advice and recommendations for improving this work, they were an important contribution to achieve successful results.

I would like to thankfully acknowledge the funding sources which made this research possible. The research project PEARL (Preparing for Extreme And Rare events in coastaL regions), funded by the European Union Seventh Framework Programme (FP7) under Grant agreement n° 603663. Also, the research project RECONECT (Regenerating ECOsystems with Nature-based solutions for hydro-meteorological risk rEduCTion), from the European Union's Horizon 2020 Research and Innovation Programme under Grant Agreement n° 776866. Moreover, this work was carried out on the Dutch national e-infrastructure with the support of SURF Cooperative. This made possible to use virtual computers to apply the optimisation framework.

There is a lot of people that contributed to build a second home so far away from home. Ceci and Javi, thanks so much for warmly receiving me in your home when I arrived at my second time in The Netherlands and for remaining very close all this time. Maria, thanks for sharing many good moments and for adapting your german punctuality for our meetings. Many thanks to Pato, Aki, Sophie, Adri, Gonza and Lari for sharing so many great experiences and the most remarkable one of being parents abroad. Also, my thanks to many friends and so nice people who were part of this path: Neiler, Angie, Fer, Juanca, Yared, Juliette, Nata, Vero, Mauri B., Jessy, Larita and Mauri M., Miguel, Pin, Mohaned, Shanoor, Mary, Irene, Natalia, Adele, Milk, Andre, Vivi, Kenichi and Jovana. My thanks to so many friends, colleagues and loved ones in Uruguay for being so close despite the distance. To Tati, Vale, Vic and Anita for your long and precious friendship making me feel as I never have left every time I visit. To my family in law for all your affection and support. Also, to the SEPS crew, for your warm welcome and farewells, I learned a lot inside this group and was there where the passion about this topic was fully developed.

And finally and most importantly, my deepest thanks to my family for their love, encouragement and help.

A mis padres, por darme las herramientas necesarias para alcanzar mis objetivos, por enseñarme a buscar soluciones por mí misma, por siempre alentarme a mejorar, por su continuo y absoluto apoyo. A mis hermanos por su compañerismo, por enseñarme tanto mientras aprendíamos juntos, por estar siempre conmigo. Además, gracias por mis excelentes cuñados y los más increíbles y adorables sobrinos. A Pablo, por todo el amor, paciencia e incondicional apoyo. Por siempre creer en mí, por compartir tantas aventuras a lo largo de estos años, y la más extraordinaria de todas: Emilia.

ABOUT THE AUTHOR

Alida obtained her BSc. in Civil Engineer, with a specialisation on Hydraulic and Environmental Engineering, in 2007 from the Faculty of Engineering, Universidad de la Republica, Montevideo, Uruguay. While studying she was already working as a consultant, task that continued until 2012. During those years she was involved in a broad range of projects related with water and environmental issues. Since 2009, she also worked for the public sector (Montevideo's Municipality), designing and implementing solutions for sewerage problems, pluvial and fluvial flooding, unplanned population growth, etc.

In 2012, looking for strengthen her knowledge and skills she started a MSc. degree in Urban Water Engineering and Management, at the Asian Institute of Technology and IHE Delft. For this she successfully acquired funds from the Joint Japan/World Bank Graduate Scholarship Program. Her Master thesis was focused on a case study in Montevideo, comparing and optimising traditional and sustainable drainage infrastructures to reduce combined sewer overflows.

In 2014, after completing her MSc studies, Alida returned to her country. There, she led a project related to reducing pluvial flooding through the application of an open detention basin. This project involved interdisciplinary work to design the detention basin as a multi-functional area. After several long meetings, the result was communicated. The result faced the opposition of technical decision makers and residents, which were not involved during the planning process. The main concern was the management of runoff at the surface, all the risks this may bring, how to maintain the proposed green spaces, uncertainties about the performance for flood mitigation. Today the detention basin exists and works properly to mitigate flooding, without multi-functionality or greenery. This was one among many learning processes which helped her to understand the importance of holistic decision-making. It was an inspiring experience and source of motivation to start the research presented in this book.

In June 2015, Alida started her PhD research at the Department of Environmental Engineering and Water Technology at IHE Delft, and the Faculty of Applied Sciences at TU Delft. Her research was part of a larger project PEARL (Preparing for extreme and rare events in coastal regions), funded under the European Union FP7. This project ended in 2018, since then her research has been funded by the project RECONECT (Regenerating ecosystems with nature-based solutions for hydro-meteorological risk reduction), from the European Union's Horizon 2020 Research and Innovation Programme.

LIST OF PUBLICATIONS

Journal Publications

Alves A., Vojinovic Z., Kapelan Z, Sanchez A., Gersonius B. "Exploring trade-offs among the multiple benefits of green-blue-grey infrastructure for urban flood mitigation", Science of The Total Environment **2019**, In Press, doi.org/10.1016/j.scitotenv.2019.134980.

Majidi, A.N.; Vojinovic, Z.; **Alves, A.**; Weesakul, S.; Sanchez, A.; Boogaard, F.; Kluck, J. Planning Nature-Based Solutions for Urban Flood Reduction and Thermal Comfort Enhancement. Sustainability **2019**, 11(22), 6361; https://doi.org/10.3390/su11226361.

Alves A., Gersonius B., Kapelan Z., Vojinovic Z., Sanchez A. "Assessing the Co-Benefits of green-blue-grey infrastructure for sustainable urban flood risk management". Journal of Environmental Management **2019**, 239, 244-254. https://doi.org/ 10.1016/j.jenvman.2019.03.036

Alves, A., Gersonius, B., Sanchez, A., Vojinovic, Z.; Kapelan, Z. Multi-criteria approach for selection of green and grey infrastructure to reduce flood risk and increase co-benefits. Water Resources and Management **2018**, 32: 2505. https://doi.org/10.1007/s11269-018-1943-3

Alves, A.; Patiño Gómez, J.; Vojinovic, Z.; Sánchez, A.; Weesakul, S. Combining Co-Benefits and Stakeholders Perceptions into Green Infrastructure Selection for Flood Risk Reduction. Environments **2018**, 5, 29. doi.org/10.3390/environments5020029

Vojinovic, Z.; Keerakamolchai, W.; Weesakul, S.; Pudar, R.; Medina, N.; **Alves, A.** Combining Ecosystem Services with Cost-Benefit Analysis for Selection of Green and Grey Infrastructure for Flood Protection in a Cultural Setting. Environments **2017**, 4, 3. https://doi.org/10.3390/environments4010003

Alves, A.; Sanchez, A.; Vojinovic, Z.; Seyoum, S.; Babel, M.; Brdjanovic, D. Evolutionary and Holistic Assessment of Green-Grey Infrastructure for CSO Reduction. Water **2016**, 8, 402. https://doi.org/10.3390/w8090402

Karavokiros, G.; Lykou, A.; Koutiva, I.; Batica, J.; Kostaridis, A.; **Alves, A.**; Makropoulos, C. Providing Evidence-Based, Intelligent Support for Flood Resilient Planning and Policy: The PEARL Knowledge Base. Water **2016**, 8, 392. https://doi.org/10.3390/w8090392

Vojinovic Z., **Alves A.**, Patiño Gómez J., Weesakul S., Keerakamolchai W., Meesuk V., Sánchez A.; Evaluation of small and large scale nature-based solutions for flood risk reduction. In preparation.

142

Conference proceedings and book chapters

Alves, A.; Sanchez, A.; Gersonius, B.; Vojinovic, Z. A Model-based Framework for Selection and Development of Multi-functional and Adaptive Strategies to Cope with Urban Floods. Procedia Engineering 12th International Conference on Hydroinformatics, HIC **2016**.

Alves, A.; Sanchez, A.; Vojinovic, Z.; Kapelan, Z. Selecting Optimal Configurations of Green and Grey Infrastructure for Flood Risk Reduction. 14th IWA/IAHR International Conference on Urban Drainage, ICUD **2017**.

Alves, A.; Sanchez, A.; Gersonius, B.; Vojinovic, Z. Selecting Multi-functional Green Infrastructure to Enhance Resilience Against Urban Floods; in: Water Security in Asia, Opportunities and Challenges in the Context of Climate Change, edited by: Babel, M.S., Haarstrick, A., Ribbe, L., Shinde, V., Dichtl, N., Springer International Publishing, **2020**.

Netherlands Research School for the
Socio-Economic and Natural Sciences of the Environment

D I P L O M A

For specialised PhD training

The Netherlands Research School for the
Socio-Economic and Natural Sciences of the Environment
(SENSE) declares that

Alida Ivana Alves Beloqui

born on 18 March 1980 in Montevideo, Uruguay

has successfully fulfilled all requirements of the
Educational Programme of SENSE.

Delft, 30 January 2020

The Chairman of the SENSE board

Prof. dr. Martin Wassen

the SENSE Director of Education

Dr. Ad van Dommelen

The SENSE Research School has been accredited by the Royal Netherlands Academy of Arts and Sciences (KNAW)

K O N I N K L I J K E N E D E R L A N D S E
A K A D E M I E V A N W E T E N S C H A P P E N

The SENSE Research School declares that **Alida Ivana Alves Beloqui** has successfully fulfilled
all requirements of the Educational PhD Programme of SENSE with a
work load of 38.6 EC, including the following activities:

SENSE PhD Courses

o   Environmental research in context (2015)
o   Introduction to R for statistical analysis (2016)
o   Research in context activity: 'Preparing and co-organizing Annual PhD Symposium of IHE
    Delft, Institute for Water Education (Delft, 2-3 October 2017)'

Selection of other PhD and Advanced MSc Courses

o   Creative Tools for Scientific Writing & Analytic Storytelling, TU Delft (2017)
o   High performance computing helping to solve water related problems, IHE Delft (2017)
o   Effective negotiation: win-win communication, TU Delft (2017)
o   The art of presenting science and Analytic storytelling, TU Delft (2017)
o   Design of infrastructures under uncertainty, TU Delft –PAO (2018)
o   Adaptive planning for infrastructure and water management & Building with Nature, TU
    Delft (2018)
o   Sustainable urban development , TU Delft and Wageningen University (2018)
o   Career development: Personal branding, presenting yourself effectively, TU Delft (2019)
o   Leadership, teamwork and group dynamics, TU Delft (2019)

Management and Didactic Skills Training

o   Supervising three MSc students with thesis (2016-2019)
o   Teaching in MSc courses 'Urban flood management and disaster risk management',
    'Advanced water transport and distribution' and 'Urban drainage and sewerage' (2017)

Oral Presentations

o   *A model-based framework for selection and development of multifunctional and
    adaptive strategies to cope with urban floods.* International Hydroinformatics
    Conference, 22-26 August 2016, Incheon, South Korea
o   *Selecting multi-functional green infrastructure to enhance resilience against urban
    floods.* Water security and climate change: challenges and opportunities in Asia, 29
    November- 1 December 2016, Bangkok, Thailand
o   *Selecting optimal configurations of green and grey infrastructure for flood risk reduction.*
    International Conference on Urban Drainage, 10-15 September 2017, Prague, Czech
    Republic

SENSE Coordinator PhD Education

Dr. Peter Vermeulen

T - #0106 - 071024 - C29 - 240/170/9 - PB - 9780367485979 - Gloss Lamination